Higher-Yielding
Human Systems
For Agriculture

HIGHER-YIELDING HUMAN SYSTEMS FOR AGRICULTURE

Edited by
William Foote Whyte
and Damon Boynton

Cornell University Press
Ithaca and London

International Standard Book Number 0-8014-1611-6
Library of Congress Catalog Card Number 83-45151

Printed in the United States of America

Librarians: Library of Congress cataloging information appears on the last page of the book.

The paper in this book is acid-free and meets the guidelines for permanence and durability of the Committee on Production Guidelines for Book Longevity of the Council on Library Resources

Contents

Preface

This book is the product of an interdisciplinary working group supported by the Rural Development Committee in Cornell University's Center for International Studies. William Foote Whyte (industrial relations and sociology) and Damon Boynton (agricultural development and horticulture) have served as coordinators of the project.

The other Cornell contributors and their disciplinary backgrounds are as follows: Randolph Barker (economics), Joseph Campbell (agricultural engineering), E. Walter Coward, Jr. (sociology), Matthew Drosdoff (soil sciences), Milton J. Esman (political science), Gilbert Levine (agricultural engineering), Robert McDowell (animal sciences), and Norman Uphoff (political science).

Two persons outside Cornell have worked closely with us, and we have been happy to include them as contributors. They are social psychologist David C. Korten, who has worked with the Ford Foundation and USAID in the Philippines, and economist Peter Hildebrand, who is with the University of Florida.

Identifying the contributors by no means exhausts the list of colleagues who have helped with this book. Other Cornell faculty members in the Colleges of Agriculture and Life Sciences, Arts and Sciences, and Art, Architecture, and Planning have given us thoughtful criticisms and suggestions on early drafts of these chapters. They have also guided us to important items in the research literature that we might otherwise have overlooked. Among these participants have been Loy Crowder (plant breeding and biometry); H. David Thurston (plant pathology); Alvin Johnson (agronomy and agricultural extension); David Lewis (regional planning); Milton Barnett (rural sociology); Davydd Greenwood, Rada Dyson-Hudson, and Billie Jean Isbell (anthropology); and Daniel Sisler (agricultural economics).

Kathleen King Whyte carried out the difficult task of editing a book initially written by many people of diverse disciplinary backgrounds. We depended particularly on her to make this an integrated and clearly written book rather than a miscellaneous set of chapters, each written in the disciplinary jargon of its author. Parts of this book grew out of a

7

state-of-the-art paper, *Participatory Approaches to Agricultural Research and Development* (Whyte, 1981), commissioned by USAID under the Rural Development Participation Project, which it supports through a cooperative agreement with the Rural Development Committee at Cornell University.

Material from *The Puebla Project: Seven Years of Experience, 1967–73* (El Batan, Mexico: CIMMYT, 1974) is used with the permission of the publisher, Centro Internacional de Mejoramiento de Maíz y Trigo. Quotations from *Caqueza: Living Rural Development,* by Hubert Zandstra, Kenneth Swanberg, Carlos Zulberti, and Barry Nestel (Ottawa: International Development Research Centre, 1979), are used with the permission of the International Development Research Centre. Material from *Small Farm Development: Understanding and Improving Farming Systems in the Humid Tropics,* by Richard R. Harwood, copyright © 1979 by the International Agricultural Development Service, is reproduced with the permission of the publisher, Westview Press.

WILLIAM FOOTE WHYTE
DAMON BOYNTON

Ithaca, New York

Part I

Foundations for
a New Approach

1

Introduction

If developing countries are to meet national food needs and alleviate rural poverty, millions of small farmers must become active participants in the agricultural research and development process. The purpose of this book is to present a new strategy for reaching these goals. We base our conclusions both on a critical look at past experience in agricultural research and development and on an analysis of emerging new systems in several developing nations.

This emerging new strategy requires a rethinking of the roles of government agencies and others who aim to stimulate and guide the agricultural research and development process. We see this rethinking as including the following major points:

(1). There must be increased emphasis upon on-farm research, with special attention given to studies and experimentation on farming systems to complement conventional programs, which have concentrated mainly on experimentation in agricultural research stations and have dealt separately with crops and animals.

(2). The new emphasis must be supported by greater interdisciplinary collaboration among professionals in the field. We foresee a much broader range of specialists working together than has been customary in the past.

(3). The new emphasis requires agricultural bureaucracies to be more responsive to the interests and needs of small farmers and to provide better coordination among the various agriculture-related agencies than currently exists. Planners need to think not only about the effectiveness of each component organization but also and particularly of ways to coordinate their activities and services so that small farmers get the help they need when they need it.

(4). The new emphasis is built upon the active participation of small farmers in the research and development process. No longer are they to be treated simply as passive recipients of what the experts decide is good for them.

This new strategy requires a basic change in how we conceptualize peasants or small farmers. If we visualize them as active participants in the change process, we necessarily abandon any notions of having experts come in simply to tell them what to do and then try to overcome resistance to change when the small farmers do not follow this guidance. We recognize that there are wide variations in the management abilities of small farmers—as indeed there are among other farmers. Small farmers should be seen as rational human beings, open to changes that would improve the lot of their families and communities and capable of weighing for themselves the potential costs and benefits of any change in their farming system.

We do not mean to suggest that small farmers always act in perfectly rational ways, but we do insist that resistance to change is no more characteristic of small farmers than it is of professionals. In fact, we will provide evidence indicating that professionals in this field have been locked into conventional strategies and organizational models even when the conventional ways of doing things were clearly counterproductive.

The new strategy calls for abandonment of the highly inefficient one-on-one relationship between the individual farmer and the extension agent in favor of dealing with organized groups of farmers. The importance of organization at the farmer level has been curiously neglected by students of agricultural research and development. After reviewing the literature, David Korten remarked: "we might expect that the difficult problems of *how* to strengthen the development role of local social organizations and otherwise involve the poor in their development would be receiving major attention in current policy documents and development journals. Yet the opposite is the case" (Korten, 1980:481).

Korten points out that a recent review of regional development by the Asian Development Bank (1977) devotes only 4 out of 489 pages to community organization. He cites another important book aiming to alleviate the world food crisis by improving small family farm production (Wortman and Cummings, 1978), which nevertheless devotes only one paragraph out of its 440 pages to farmer associations or cooperatives. In various other documents and even current development journals he finds similar neglect of farmer associations, cooperatives, or any form of community organization.

We assume that there are important economies of scale when buying and selling that are beyond the reach of the individual small farm family. Those economies of scale can be achieved through an effective organization of small farmers—but "effective" is the key word, for it is probably true that more farmer cooperatives fail than succeed.

Nevertheless, the potential for improvement through organization is important enough to deserve far more attention than it has yet received in publications on agricultural research and development. The connection between farmer organization, devoted to the improvement of the economic well-being of its members, and political influence may be feared by some, but we should see the value in having more rural influence upon government agricultural policies and programs, to get them adequately funded and supported and to orient them to real development needs.

The need for interdisciplinary collaboration in agricultural research and development has received recognition but in practice is often lacking in programs in the field and even in the publications of students of agricultural research and development. Such interdisciplinary collaboration as exists is usually found in the biological sciences of plant pathology, plant breeding, entomology, agronomy, and the like. We see beginnings of integration of the social sciences into such programs, but progress has been slow, and the discipline most often integrated has been economics. It is only recently that professionals trained in social anthropology and sociology have begun to find their way into international agricultural research centers and national agricultural research and development programs. One of the aims of this book is to show how interdisciplinary collaboration can be broadened and strengthened.

Who Are Small Farmers?

It seems reasonable to begin with a definition of the small farmer, yet there is no precise definition that satisfies us fully. We can start by saying that the small farmer is an individual who works a small parcel of land. Even here we run into problems, however, because there is no fixed upper limit for what constitutes a "small" holding. The farmer who is working two hectares of rain-fed land and has only one crop season per year would be a small farmer by almost any definition, but consider the farmer who has two hectares of irrigated land in a part of the world where he can fit two or three crop seasons into the same year. The farms in both cases are equally small, but we are likely to find that the first farmer has to struggle to rise above the subsistence level whereas the second farmer buys and sells in the market and is relatively prosperous.

The lower limits of the population addressed in this book are clear. We do not deal here with people who own or rent no agricultural land. There are millions of landless laborers around the world, and their problems deserve the attention of social scientists and policy planners.

But our research and experience are most relevant to those farm families growing crops and raising livestock for their own use or for sale so that they profit or lose directly depending on the success of their agriculture. If the landowner with hired laborers achieves increased agricultural production, he may raise their pay, but this does not necessarily follow. The owner may keep all the benefits himself. In any case, the outcome for laborers does not simply depend on the productivity of the land and the selling price of the produce. Their income will also depend upon the relationship between the supply and demand for labor, the extent to which laborers are organized to protect their own interests, and government policies and programs that support or undermine the interests of the landless. The interrelations among variables affecting the income of agricultural laborers are sufficiently complex to require special treatment beyond the scope of this book.[1]

We include in our consideration here tenants and sharecroppers who work land for themselves, even though they share some of their income or crops with the owner. Of course, their economic welfare depends not only upon their agricultural success but also upon the nature of the formal or informal contract with the owner, as well as upon their earnings from supplementary or even primary outside work. Still, if practitioners of agricultural research and development devise improved systems to aid small farmers, tenants and sharecroppers are likely to benefit from these efforts, and therefore they fall within the scope of our concern.

Our attention is not limited to those families that grow all or nearly all of their own food and in which all active members work on the family farm. Around the world, we see millions of farms that are so small or suffer such ecological handicaps that it would be impossible for the family to feed all of its members from the farm produce, no matter how hard they work or how much they learn from the agricultural sciences. The survival of such families will continue to depend upon the earnings of one or more members as farm laborers or workers in nearby towns or cities.

Such people may indeed benefit from advances in the agricultural sciences that enable them to increase their yields. If a proposed innovation requires more money and more labor, however, they may not be able to afford to make the change, even with the prospect of increased yields and farm income. They must judge whether they more urgently need the income and security from family members currently employed off the farm.

[1]For an interdisciplinary study of the landless and near-landless, see Esman (1978).

Given the varied economic and ecological conditions around the world, it would be unwise to specify any particular size of farm or other conditions beyond which the small farmer must be considered a medium or large farmer. Let us say simply that we are particularly concerned with those Third World farmers who so far have been largely bypassed by the great advances of agricultural sciences of recent decades. More specifically, we will focus upon farmers who face severe handicaps in size of land holdings, ecological conditions, and the resources available to them. A term used by some of our Cornell colleagues—*limited-resource farmers*—is perhaps more exact than the commonly used term, *small farmers* (though we will continue to use the latter because it is less cumbersome). In essence, the aim of this book is to devise ways of bringing disadvantaged farm families more effectively into the stream of advancing agricultural technology.

Plan of the Book

Although we focus primarily upon the present and upon what we project for the future, any attempts to sketch a new model of agricultural research and development must be based to a considerable extent upon our understanding of past and current models. We therefore begin in Part I (Foundations for a New Approach) with a review of past research and development efforts, indicating how increasing recognition of the limitations of these programs has led to change.

In part II (Physical and Biological Bases for Small Farm Development), we examine the complex range of nonhuman elements that must be integrated in the creating of efficient farming systems. As the title of this book suggests, the interactions and activities of human beings must be organized into systems of socioeconomic and even political relations if they are to create efficient farming systems. Thus in Part III (Social Systems from Farm Families to National Programs), we present a way of thinking designed for understanding the role of humans in building efficient systems at village, regional, and national levels.

In Part IV (Organizational Implications), we discuss community and farm organization and examine the nature and problems of government agencies involved in agricultural and rural research and development.

In Part V (Implications for Research, Education, and Government Policy), we consider implications of all the foregoing for research in the plant and animal sciences and in the social sciences. We also discuss the need for reorientation of education, and, finally, we consider government policies required for supporting the emerging participatory systems of agricultural research and development (R&D).

We cannot expect to advance the interests of small farmers and the rural poor simply by channeling more money into agricultural research and development, by training and hiring more high-level professionals, and by tinkering with traditional styles of organizing and administering such programs. Our diagnosis points to the need for a profound rethinking of the nature of research in the plant, animal, and soil sciences and also in the social sciences. If future researchers and practitioners are to meet the challenge posed by the emerging models of agricultural R&D, they need an education that does not require them to begin their professional careers by having to overcome some of the traditional ways of thinking and acting.

We do not claim that many of the ideas presented here are original with the present authors. We have drawn freely upon the contributions of fellow researchers in Asia, Africa, and Latin America, but so far the evidence required to describe the emerging trend in agricultural R&D has been scattered about in many sources. Here we aim to pull the information together and systematically describe the nature of the new pattern of research and development. We recognize that this new pattern has been emerging so recently that we lack adequate information on which to evaluate performance and promise. Nevertheless, students, planners, and practitioners are showing a growing interest in this new way of organizing R&D, and it is important to provide the best guidance we can offer on how to turn potentialities into realities.

Although focusing upon an emerging reorientation, we are not arguing that research directors should abandon disciplinary specialization and monocultural projects to breed improved varieties of particular plants. We are simply proposing to balance established ways of doing research—which have yielded extraordinary gains in improved plant varieties and in other biological aspects—with development of interdisciplinary projects in which small farmers are active participants, along with researchers and extensionists.

2

Research and Development
Organizations: A Reevaluation

Now that growth theories that once were widely accepted and that confidently predicted broad-based improvements in living standards in less developed countries are in question, practitioners, researchers, and scholars in all the disciplines have been groping toward new definitions and new approaches to development. From this experience a fresh consensus is emerging: that economic development and technological progress must be designed and implemented so as to combine growth and equity, and that rural development strategies must begin to yield direct benefits to the great majority of rural people. This new consensus has important implications for research and development organizations.

Exporting Familiar Models

Specialists operating outside their homeland tend to "export" familiar organizational models. Former French colonies still have the familiar apparatus of French public administration, from police systems to *écoles nationales d'administration*. In former British colonies the structural models of British universities survive as reminders of the colonial experience, along with common law and the English language. The United States attempted to replicate American models of public education and of its civil service system in the Philippines. The reasons for these carryovers are not hard to find. During the colonial period, institutional transfer combined the power and prestige of the metropolitan country with the experience of its specialists. These were the models that expatriate officials were familiar with, had confidence in, and believed would bring significant benefit to people in the colonies.

In the post–World War II occupation of Japan, the drafting of the present Japanese constitution involved similar efforts to transfer the institutions and organizational forms of the victorious power, from the California system for selecting judges to parent-teacher associations. Though Japanese institutions were too firmly established and the six-year occupation period was too brief to allow most of these institutional transfers to survive, many of their influences remain. When large-scale foreign aid was initiated first to Latin America during World War II and then to the newly independent former colonies of Africa and Asia, United States organizational forms followed. Whether in teacher training, highway departments, or public budgeting, Americans took their organizational models with them. U.S. experts were prominent in United Nations technical assistance activities, U.S. technicians served in successive bilateral aid programs, and U.S. advisers represented the major American foundations overseas.

In no other field were U.S. organizational models so eagerly transferred as in agriculture. Immediately after World War II the United States was the dominant military and economic power. The prestige of U.S. industry was exceeded only by the prestige of its agriculture. U.S. agriculture had become so fantastically productive that it was able to avert famine among both victors and vanquished. It was universally acclaimed as the most successful, bountiful agricultural system that the world had ever seen. Leaders of foreign countries were as eager to emulate it and to enjoy its benefits as Americans were to share their knowledge. Students and admiring observers came from all corners of the world, even including Khruschev in 1959, to see and to learn the miracles of U.S. agricultural science and the institutions that made it work. American administrators and specialists thus felt justified, indeed morally obligated, to make this experience available to the less fortunate peoples of the world. Since most agricultural technologies are not protected by patents, there seemed to be no barriers to the rapid diffusion of U.S. agricultural know-how throughout the' developing world.

The Research Institute Focusing on a Single Crop

Before the period of U.S. influence in less developed countries, the European powers developed in some colonies large, interdisciplinary research institutes devoted to a single, high-value export crop. These institutes were usually staffed by expatriate scientists; their clients were for the most part expatriate-owned estates and plantations; financing was often shared by the planters and the colonial government. Although successful within a limited sphere, they functioned as enclaves within the colony and were unconcerned with indigenous agriculture except where local growers were integrated into and closely regulated by the system of export crop cultivation.

They were seldom involved with subsistence food crops grown and consumed by the local peasantry and low-income town dwellers. They were oriented primarily to improving the quality, increasing the yields, combating diseases, and lowering the costs of production for export crops. Production, processing, and marketing were controlled by colonial planters and corporations, often with the assistance of government-managed marketing boards. A very successful example was the Malayan Rubber Research Institute. Only since independence have serious efforts been undertaken to extend the research and services of these institutes to meet the needs of smaller local producers of major export crops.

The work of these specialized institutes was often of high technical quality, but they made little effort to involve local personnel in professional or managerial roles, and they had no impact on the great majority of peasants. They were high-technology enclaves with few backward or forward linkages to indigenous agriculture.

Extension Services and Supervised Credit

Agricultural extension was the first organizational form introduced to less developed countries by U.S. specialists. The premise was that an enormous storehouse of knowledge on improved agricultural practices had been developed in the United States over many years and was available to be drawn on for the benefit of farmers in underdeveloped countries, who were assumed to be eager for this information. Extension services were seen as the means to transfer technology quickly from experiment station to farmers in the United States and from the United States to farmers abroad. The underlying assumption of Point IV,[1] beginning in 1949, was that the transfer of "Yankee know-how" would rapidly increase production and improve living conditions both of farmers and of underfed urban consumers in the poor countries of the world, at the same time demonstrating the moral and pragmatic superiority of the "free world" in the emerging struggle against international communism. The approach was to transfer existing knowledge quickly, not to generate new knowledge or practices, assuming that the existing knowledge would be similarly productive in new environments. It was no accident that the first director of Point IV, Henry Bennett, was a product and an enthusiastic proponent of the U.S. pattern of agricultural extension.

During the decade preceding 1955, versions of the American agri-

[1]Inaugural Address of President Truman in 1949: "Fourth, we must embark on a bold new program for making the benefits of our scientific advances and industrial progress available for the improvement and growth of underdeveloped areas."

cultural extension service, complete with 4-H clubs and home demonstration agents, were installed in every Latin American country (Mosher, 1957) and in many Asian countries where the United States maintained a foreign assistance presence. (There was little U.S. activity in Africa during this period preceding decolonization, except in Liberia and Ethiopia, where agricultural extension services were initiated.) Through the well-tested methods of information transfer—training and demonstration—farmers would be made aware of better agricultural and livestock practices. According to U.S. doctrine, the functions of extension were to educate farmers and their families in improved practices and increase their ability to make sound decisions. Extension services were to avoid supply, marketing, and regulatory activities, which were the responsibilities of other public and private agencies and would distract extension personnel from their educational functions.

During this early period, another organizational form that competed (unsuccessfully) with the extension approach was the supervised credit system drawn from the experience of the Farm Security Administration (FSA). The FSA was one of a succession of New Deal efforts to assist tenants, sharecroppers, and marginal farmers—the rural poor—in the United States. The premise was that poor farmers lack not only information about modern and efficient farm practices but also access to production inputs—seeds, fertilizers, and equipment. The main instrument adopted to help these farmers was supervised credit, tied to detailed farm management plans, which would indicate how the inputs purchased with low-interest loans should be used.

Beginning with the agricultural and industrial booms immediately before World War II a fundamental clash developed between the more influential members of the U.S. agricultural establishment, who proclaimed the priority of maximizing production, and the advocates of small farmer strategies, who fought for the preservation of the "family farm" on political and social equity grounds. The former argued that instead of wasting resources on futile efforts to make marginal farms viable, farmers on poor soils or with insufficient holdings or limited access to production inputs should be encouraged to move to industry, where good jobs awaited them at incomes far above what they could earn as marginal farmers. Agriculture, it was argued, was becoming increasingly commercialized, scientific, and mechanized, thus requiring large-scale operations; farmers who could not meet the test of competitive efficiency should give up farming. Their holdings would be consolidated into larger units.

In this contest, social Darwinism won an easy victory. The FSA became the stepchild of the U.S. Department of Agriculture (USDA). Domestic programs to help marginal farmers were underfinanced, and government efforts were concentrated on assisting commercial agriculture to become ever more productive and profitable. In contrast with the extension service, whose "extension" abroad was strongly supported by the agricultural establishment in the USDA and the major farm organizations, any proponents of overseas programs focusing on small farmers were disadvantaged by the lack of a strong domestic support base in the United States.

A few programs oriented to small farmers did survive, along with credit unions, community development, and similar efforts to reach the rural poor. These programs tended to operate on a small scale, often with the participation of private voluntary agencies and modest assistance from government. The main effort was concentrated on agricultural extension services, for this model commanded an influential domestic constituency and was associated with the formidable production achievements of United States agriculture. The extension model tended, however, to pay little attention to structural constraints on improved agricultural practices. There was no indigenous United States model for land reform. Even the defeat of the Confederacy had not resulted in land reform. To be sure, the United States did facilitate radical and successful land redistribution in Japan, Korea, and Taiwan following World War II. But with the onset of McCarthyism and the Eisenhower administration beginning in 1953, land reform was considered an inappropriate and unacceptable interference with property rights. Milton Esman recalls a meeting with the administrator of the International Cooperation Administration (ICA), the U.S. bilateral aid program, in 1956 in which U.S. assistance to land reform in Vietnam was emphatically rejected, despite the political benefit that might have accrued to the embattled U.S.-supported Diem regime. With the Alliance for Progress[2] during the early days of the Kennedy administration, land reform became a popular cause, but no serious efforts were made to implement it in the absence of support from Latin American governments and ambiguous approval in the U.S. Congress.

The Land Grant University

Beginning in 1954 the extension approach was superseded by and incorporated into a more comprehensive organizational structure for agricultural development, one with very powerful influence both in the

[2]Charter of Punto de Este, Uruguay, 1962, as expounded by President John F. Kennedy.

United States and abroad—the U.S. land grant university. The adoption and promotion for two decades of the land grant university model as the principal organizational vehicle for agricultural research and development was the result of a learning process. During the first decade or so of foreign assistance it became clear that much of the knowledge contained in the "storehouse" of U.S. agricultural practice was not readily transferable to most developing countries because of differences in crops, soils, and climate, in the resources available to local cultivators, and in the institutions and public services associated with agriculture. There were not enough agricultural scientists and other specialists to staff extension services, which were in any event underfinanced. Nor were there adequate educational facilities to train agents properly either in technical or communications skills. Moreover, the extension services often had little of interest to impart to local farmers (Rice, 1971).

It became clear that agricultural progress in less developed countries could not be achieved through a short-term process of transferring known methods and practices. Sustained agricultural development would require not only individual foreign aid technicians working on small projects with a single local counterpart but teams of specialists working with local groups to build appropriate institutions. Decades might be needed to build the necessary research, educational, and extension institutions and to innovate new technologies suitable to the specific agronomic, cultural, and institutional problems of agriculture in a variety of usually tropical environments. Specifically, it would be necessary to undertake research on improved crops and cultivation practices suitable to specific local environments, to train large numbers of specialists in all the agricultural sciences, and to reinforce extension—the diffusion of knowledge—with an adequate institutional base for the generation of useful information and practices. All of these would depend on adequate educational facilities.

The logical model was the U.S. land grant university, which combined education, research, and extension, each reinforcing the other, within a single institution. Unlike elitist faculties of agriculture inspired by European experience, research was to be guided by the practical needs of farmers as communicated through extension agents. Teaching was to be informed both by the latest research findings and by knowledge of the conditions and needs of farmers. Extension would benefit from the latest findings of laboratories and experiment stations. The land grant doctrine seemed to be especially relevant to the needs of developing countries because it emphasized service to the farmer (as opposed to education for its own sake or for the production of elites alienated from their farm constituencies) and the application of science to everyday needs. Both of these concepts represented radical but necessary innovations in agricultural education in nearly all developing countries.

Even more than the extension service, the land grant institution (which incorporated the extension function) was the organization most dynamically associated with America's agricultural achievements. Once it became clear that the United States was prepared to make substantial expenditures and long-term commitments on behalf of agricultural development overseas, the land grant model became the obvious organizational vehicle. It was demanded as much by foreign officials and agriculturalists, many of them U.S.-trained, as it was promoted by U.S. educators, scientists, and practitioners deeply convinced of its universal utility as a democratic instrument for serving agriculture. Technological progress and democratic goals were thought to be inherently compatible. Little attention was paid to the structural inhibitions to agricultural modernization in developing countries or to the relative distribution of benefits and costs. If problems arose, they could be solved.

The land grant model fitted conveniently into the domestic political strategies of those responsible for American aid programs. With the successful termination of the Marshall plan, which had facilitated the rapid reconstruction of European economies and was credited with turning back communist expansion in Western Europe, it was necessary to construct a new domestic constituency for foreign aid as a long-term program. The improbable coalition of militant anticommunists (who supported foreign aid so long as it promised to contain communism in the Third World) and humanitarians (who believed that the United States had a moral obligation to help the world's poor) needed reinforcement. Harold Stassen, the administrator of the Mutual Security Administration, recalled from his days as governor of Minnesota the influence of that state's university in state politics. If the state land grant colleges could be associated with the foreign aid program, not only as sympathizers but as active participants, their considerable political influence could be mobilized in the annual confrontations with Congress for budgetary support. Under Stassen's leadership the land grand universities became actively involved as agents for implementing United States technical assistance programs abroad. A large proportion of their effort was devoted to agriculture and to building educational institutions overseas. As a result, impressive land grant–style universities now flourish in many developing countries, providing well-trained graduates in all the agricultural sciences to staff expanding networks of research, extension, and other agricultural public services and to work in agrobusiness enterprises.

Unfortunately, several elements that contributed to the success of the land grant university in the United States are absent in most developing countries. A primary element is the availability to the majority of farmers of complementary services both from government and private sources. The U.S. farmer can draw on multiple sources of production and marketing information. He is literate and numerate and, with access to many sources of information, he can consider adaptations that enable him to take advantage of new opportunities and adjust to changing conditions. He has access to credit from several agencies, some of them supported and underwritten by government. These funds, available at reasonable interest rates, enable him to purchase equipment and production inputs from private dealers or cooperatives and thus take advantage of the most recent and profitable technological innovations. He has available a network of storage, processing, transportation, and marketing services which enables him to dispose of his output efficiently. Crop insurance on reasonable terms and price supports for some crops provide both incentives to production and reasonable security against some of the risks that farmers face.

These networks of service available to the U.S. farmer are in marked contrast with the conditions under which all but a small minority of peasants operate in most developing countries. Cooperative extension units in the United States can usually recruit and pay trained professional extension officers, not to mention home demonstration agents, in sufficient numbers to work with individual farm families on a one-to-one basis. Most developing countries cannot afford this luxury. The ratio of extension agents to farm families is often so low that they can reach only a small proportion of the cultivators in their service area, and these usually turn out to be the larger, more prosperous, and therefore more influential farmers (Leonard, 1977). Extension agents frequently lack vehicles and fuel to travel in their own areas even when adequate roads are available (Axinn and Thorat, 1972). U.S. extension doctrine, which rejects involvement in supply and marketing functions, can be of little use to small farmers on marginal soils who lack access to inputs on reasonable terms and to effective processing and marketing facilities.

In the United States farmers have political clout. The county governments to which they pay taxes share in financing the cooperative extension services. The county agents work for them. Farmers are organized into powerful, active, and very articulate interest groups. As individuals or through organized groups, U.S. farmers have been involved in planning and in actively responding to research and extension activities. They are not considered by agricultural administrators or technicians to be socially inferior. The majority of university teachers, researchers, and extension agents in agriculture are the children or siblings of practicing farmers. Farmers' needs and priorities are communicated quickly, accurately, and decisively to responsive agents of the land grant system. In most developing countries, none of these conditions prevail for the great majority of farm families. They are low-status clients and not the employers of extension personnel; they are not organized for advocacy purposes; and they have little ability to articulate their needs effectively or to influence government agencies.

Moreover, U.S. agricultural policy and the land grant system have increasingly emphasized "efficient" production for commercially oriented farms (McConnell, 1969). Many millions of American farmers have transferred out of agriculture to industries and service occupations. The "family farm" has been redefined to encompass much larger acreage and greater capitalization. Agricultural public services, including research, have increasingly concentrated on large-scale commercial farms and agrobusiness enterprises. Though long-term social costs may be high, these priorities are feasible in a country where land and capital are relatively plentiful, labor is scarce and expensive, and jobs in industry are readily available.

Like cultivation practices and crop varieties based entirely on experience with temperate climate and soils, policies developed in the United States have not fitted the socioeconomic environment of most developing countries, where the majority of rural households are headed by wage laborers, small tenants, and marginal farmers (Esman, 1978). Where land and capital are scarce and expensive, efficiency should be measured by yield per acre rather than yield per worker. Nevertheless, the criteria of efficiency derived from recent U.S. agronomic and socioeconomic conditions have been carried over to developing countries, where the agricultural environment is radically different. The tacit assumptions have been that farmers in general would have access to the inputs and services that would make scientific farming possible and that displaced farmers could find employment in other lines of work. Thus studies of rural development in the 1950 to 1970 period paid very little attention to problems of involving small farmers and the rural poor (Propp, 1968; Beers, 1971).

Like the more limited extension system, the land grant model was predicated on the concept of a wide outreach from the service-providing institutions to its rank-and-file clientele. Yet in most developing countries, the prime beneficiaries have been the minority of substantial farmers who can be readily contacted by understaffed extension organizations and can run the risks of innovation and afford the seeds, fertilizers, pesticides, and mechanical equipment usually associated with progressive farming. Only recently have some of these institutions begun to take employment, welfare, and distribution of income explicitly into account, to experiment with using paraprofessionals and other ways of reaching marginal farmers, or to work on intensifying cultivation on very small holdings and increasing earnings for the landless or near landless.

The International Agricultural Research Centers

The research and development model for agriculture established most recently with strong United States influence has been the international agricultural center (Consultative Group on International Agricultural Research, 1976b). The impetus for these centers, now fourteen in number with a total annual budget of over $163 million, was the urgent need that appeared in the late 1950s to head off impending famine resulting from rapid population growth and lagging production of foodstuffs in most of Asia, Africa, and Latin America. There was convincing evidence that major increases in food production were technically possible if sufficient energies, skills, and resources could be concentrated on finding solutions to problems under the agronomic conditions of the Third World. A model for such an effort may have been the Manhattan project, which demonstrated, in producing the atomic bomb, what multidisciplinary teams of scientists could do when turned loose on a specific problem. Closer to home was the encouraging experience with the Rockefeller Foundation program begun in Mexico in 1943, which by the late 1950s had demonstrated promising results with hybrid corn and dwarf wheat varieties suitable to nontemperate climates. The first international agricultural center was organized by the Rockefeller and Ford foundations, which jointly established the International Rice Research Institute (IRRI) in the Philippines in 1960.

The elements of this "big science" model for agricultural research and development are (1) a critical mass of highly qualified scientists operating in interdisciplinary teams, (2) adequate and secure funding, free from budgetary anxieties and unreasonable pressures for quick results, (3) a nonpolitical, scientific environment managed by scientifically qualified administrators, and (4) a clear mission toward which energies and activities could be directed. The original leadership came from plant breeders and plant pathologists, who had achieved eminence among the agricultural and biological scientists similar to that of nuclear physicists in the physical sciences. In recognition of these achievements, Norman Borlaug, long associated with the wheat program of the Centro Internacional de Mejoramiento de Maiz y Trigo (CIMMYT), was awarded the Nobel Peace Prize in 1970. The centers conceived their task to be the innovation of improved technologies. It was thought that once these were available, ways would be found to get them out to the field where farmers would adopt them as eagerly as the Iowa farmers took to hybrid corn in the 1930s. The emphasis was on high-yielding varieties under optimal conditions, especially the availability of irrigation, fertilizers, and chemical pesticides. In the early years centers were staffed almost entirely by natural scientists.

The early successes of the centers were spectacular, exceeding even the optimistic expectations of their founders and of the scientists themselves. IR-8, a new rice variety innovated by IRRI just a few years after its establishment, and its successor varieties are now widely diffused and adopted; they have helped to avert famine in many rice-eating areas. Though the plant breeding efforts continue and new pest- and disease-resistant, high- or low-temperature-tolerant, and drought-resistant varieties continue to be produced, the centers have found it necessary to pay more attention to the diffusion of the new varieties and their adaptation to specific microenvironments. This effort has necessitated the training of scientists from cooperating countries to staff local research stations, the development of outreach programs, including packages of practices to facilitate adoption, and cooperative R&D programs with national agricultural research institutions. The international institutes must depend on national capabilities to do adaptive research and on national extension services to diffuse the new varieties to local farmers. The difficulties experienced in forging these links led to the establishment in 1980, under the auspices of the Consultative Group on International Agricultural Research (CGIAR), of the International Service for National Agricultural Research (ISNAR). The new service provides advice and technical assistance to governments interested in strengthening their agricultural research and extension capabilities.

As leaders of the international agricultural research centers began to look beyond the spectacular increases in national production of basic grains, they came to recognize that poor farmers, without access to irrigation or the resources to buy the recommended inputs, had been largely bypassed by the progress of the Green Revolution. Policy makers began to devote more attention and resources to the problems of these limited-resource farmers. To gain a better appreciation of the needs of such farmers, the centers are adding small numbers of social scientists to their staffs and are attempting other ways to provide closer linkages with national institutions concerned with small farmers (McDowell, 1981).

Benefits to these less favored farmers in developing countries must come more slowly than the first breakthroughs of the high-yielding varieties because few such farmers enjoy the agronomic and economic conditions that made the earlier advances possible. There will have to be much greater emphasis on farming systems, on multiple cropping and interactions with livestock, and on using locally available resources rather than costly commercial inputs. These constraints greatly complicate the task of the scientists. Unlike the United States, where marginal farmers could abandon farming and move to better paying industrial jobs in the cities, in most poor countries the population of rural areas is increasing despite large-scale migration to the cities. There are not enough nonagricultural jobs to employ the growing rural labor force at current rates of population growth. Therefore, the rural areas of many countries will have to provide employment for even larger numbers, while the average size of holdings continues to decline. The centers have recently begun to address more of their attention to this large and growing number of very small and marginal farmers and to the landless workers. It is to this growing and disadvantaged group that we are addressing our efforts as well.

In contrast to the European models, extension services, the more comprehensive land grant institutions, and the international research centers have been primarily interested in food crops, in improving indigenous agriculture, and in helping all classes of farmers. These objectives represent major innovations in most developing countries. Yet for reasons we have already discussed, these models have not been successful in benefiting the majority of cultivators. The one U.S. model, supervised credit, which was specifically oriented to marginal farmers, was never given priority in U.S. foreign aid.

All of these institutions have represented "top-down" processes of technological innovation. Scientists define specific problems, innovate and test new technologies and improved methods in laboratories and research stations, and turn them over to extension organizations for diffusion. The farmer's role is essentially passive; his knowledge and experience are not considered to be necessary inputs into the research process. He has no direct influence on the choice or definition of problems to be studied, the testing of innovations, or the criteria for evaluating results. If he declines to adopt new and supposedly better technology, the tendency has been to deplore his inherent convervatism or the incompetence of extension staffs. Only recently have efforts begun to orient substantial research and development efforts to the needs and circumstances of small and marginal farmers. Attempts to involve them in research and development have lagged even further, though farmers in the United States have been involved from the beginning of the cooperative extension system and land grant institutions, helping to identify problems for research and passing judgments on the results. How that necessary transformation can be brought about in developing countries is the major theme of this book.

Four Organizational Models for Agricultural Research and Development

We can identify three major organizational models for agricultural research and development which have been inspired by European and United States experience. An emergent fourth form may be particularly well suited to the needs of small farmers in developing countries.

The first is the *enclave,* represented by the research institutes that were common in European colonies. These institutions focused intensively on a single high-value export crop. Scientists addressed their efforts to foreign-owned plantations or to local small holders integrated into the export economy. The flow of information was vertical, from the research institution down to relatively few users. There was no intention to serve the great majority of local farmers.

The second is the *diffusion model,* represented by U.S. extension and land grant universities. The intention was to apply science to the practical needs of all classes of local farmers, including those who grow food. These models were explicitly democratic and were intended to make research results widely available through the spread of information and improved practices. For unforeseen reasons, this diffusion has not happened. Benefits have flowed primarily to those better endowed with land, capital, education, and influence.

The third is *big science,* which is in some respects a hybrid between the enclave and diffusion models. It incorporates the high-powered scientific research center of the enclave with the extension-outreach systems of the diffusion model. The research scientists are in command, defining problems and evaluating results. They seek to disseminate results as widely as possible through existing extension institutions in cooperating countries. They encounter the same limitations as to who can be reached and who can adopt the innovations as do the diffusion institutions. They have begun to search for new ways to address the needs of the majority of small and marginal farmers, as have many universities and extension systems in developing countries.

A number of research centers are beginning to experiment with a fourth or *interactive model* to meet the need for a new organizational form to benefit small farmers. Many of the same elements present in earlier models are included, along with a new factor that appears increasingly crucial: farmer organizations, formal and informal. Farmers, including small and marginal cultivators, are incorporated actively into the process of research and development; scientists tap their knowledge and experience in defining problems, identifying constraints, testing innovations, and judging results. The farmers then draw on local managerial ability, organizational resources, and contacts to apply improvements they find useful to their farming systems. Such innovations must fit into the cropping and livestock systems of small and marginal farmers and take their specific needs, constraints, and capabilities fully into account.

This book examines efforts to reorient agricultural R&D to create more productive systems or organization that encompass small farmers and their families as well as scientists and technicians.

3

Major Agricultural Research
and Development Projects:
A Reevaluation

In this chapter we review four important regional projects involving agricultural research and development: Comilla in East Pakistan (now Bangladesh); Chilalo Agricultural Development Unit (CADU) in Ethiopia; Puebla in Mexico; and Caqueza in Colombia. Each of these projects was an important attempt to integrate a number of development efforts in a focused geographical area. We will concentrate on the lessons that can be drawn from successes and failures of these projects.

We devote more attention to Puebla and Caqueza because the lessons they produced are affecting the shape of new R&D strategies, especially in Latin America. The designers and leaders of Caqueza had substantial exposure to Puebla and have made clear what they learned from Puebla's successes and limitations. Through linking the Puebla experience to that of Caqueza, and both projects to later emerging national programs, we are able to trace the evolution of ideas about new models of agricultural research and development.

Comilla

Comilla was the earliest of the four projects, organized in 1959 with support from the Ford Foundation and the government of Pakistan (Blair, 1974). The project area was located in a *thana* (subdistrict) with a population of two hundred thousand crowded into an area of approximately one hundred square miles. The average landholding of farmers in East Pakistan was 1.5 hectares; in Comilla the average was .7 hectares, and there was less variation around this mean than elsewhere in the region. Thus planners started with an area very poor in land per family but relatively homogeneous.

Rice was the principal crop in Comilla. Where an effective system of irrigation was available, farmers could grow up to three crops of rice a year, but generally the poor conditions of irrigation limited them to two.

The project began with a nine-month period for overseas study and planning by members of its entirely Pakistani staff. During this time at Michigan State University, the Comilla staff was fully exposed to the development ideas available in the United States, but basic decisions remained under the control of the Pakistanis.

They agreed on three principles that appear remarkably modern in present-day agricultural R&D thinking:

First, they would establish locally an academy for research on agricultural and community development and for the training of local leaders.

Second, research designed to discover the interests and needs of the farmers and to understand their agricultural practices was to be carried out before any decision regarding interventions was made.

Finally, Comilla's farmers would be active participants in decision making and in the total development process.

The research highlighted important biophysical and socioeconomic problems. The Comilla area was subject to periodic floods and drought. Small farmers had no access to credit. There was no effective extension service to provide them with information and ideas.

The project first organized credit cooperatives. The project staff served as intermediaries between the banking system, which provided the credit in a lump sum, and the credit cooperatives. These cooperatives received the credit for their members and thereby assumed a collective responsibility for repayment. Comilla staff members then worked with the credit cooperatives to make them channels for information on farming and marketing and also conduits for the inputs that the farmers would otherwise have had to purchase individually. Each credit cooperative elected a manager for credit and collections and a model farmer for conveying technical knowledge to fellow members. These paraprofessionals received training at the academy one day per week.

The results were impressive, both in increased yields of rice and in organizational activity. In the Comilla area generally, between the crop years 1962–1963 and 1970–1971, rice yields increased by 20 percent. In a shorter period, 1963–1964 to 1969–1970, cooperative members increased their yields by nearly 100 percent.

At the start of the project, only about 1 percent of the land in Comilla had dependable irrigation. The academy organized a public works project to improve and extend the irrigation system. The academy also organized project committees and provided management training in planning, implementation, and fiscal control. Local people working on the irrigation project were compensated with wheat purchased by the government of Pakistan in local currency from the United States under its Public Law 480 program. During the first year, forty-five thousand man-days of labor resulted in saving six thousand acres of rice land from flooding.

As a means of increasing rice production during the winter dry season, the academy stimulated the formation of pump groups, each of which contributed its labor and was provided with a low-cost diesel pump and subsidies for fuel, fertilizers, and seeds. The manager of the pump group and the pump operator also received training at the academy once a week.

By 1973, with Comilla extending beyond the orginal *thana*, 32,924 pumps were irrigating slightly over half a million hectares. Planting a third crop now became possible for almost 850,000 farmers, and the dry-season crop production rose from 830,000 tons in 1966–1967 to more than 2 million tons in 1972–1973 (Blair, 1974:71).

Impressed by the success of the Comilla project, government officials were determined to extend it to the entire province. Academy officials argued that expansion should come slowly in stages as the academy developed the people and organizations to manage the increasing responsibilities. As too often happens, government officials were determined to go in one jump from a pilot project to a massive application of the new model. Furthermore, they did not extend the entire model but emphasized the pump groups, leaving out the research base and the capacity to train and supervise this greatly expanded organization.

The results should have been predictable. Later research found that the local elites had taken control of most of the new pump groups and had managed to take the lion's share of the benefits for themselves. On the average, pump group managers owned three times as much land as the members, and members serving on the managing committee for the pump groups received fifteen times as much money in loans as the rank and file. Nor did this distribution of loans reflect the presumed creditworthiness of the borrowers. Harry W. Blair (1974:58) found a direct correlation between the size of the loans and failure to repay: "The maldistribution of both loans and overdues [at Comilla] . . . is the result of the same factor: control of the cooperative structure by the large farmers, who appear to be tied into the traditional leadership structure and who stay in power year after year, despite the requirement for yearly election of a manager and a model farmer" (Blair, 1974:60; see also 1978).

The spread and use of the technological innovation foundered on the rocks of social stratification and political influence. The agricultural

R&D effort could not be carried through without regard for the ways communities were structured and opportunities allocated.

CADU

The Chilalo Agricultural Development Unit (CADU) began in 1967 as a joint project of the Swedish International Development Authority (SIDA) and the Imperial Ethiopian Government (Cohen, 1975). The project emerged out of the relationship between Sweden and Ethiopia, begun in the 1860s by missionaries and later reinforced by various government projects, focusing first on communication, education, and health.

CADU was intended to benefit subsistence and tenant farmers of Chilalo Awraja (subprovince). Two years of agricultural and economic studies led to the selection of Chilalo as the project area. In the beginning, all of the top-level professional staff were expatriates, making this the most foreign-dominated project among the four here considered. Ethiopia was far behind the other project countries (Pakistan, Mexico, and Colombia) in the availability of national professionals in agriculture, however, so the project was possible only with a heavy infusion of foreign professionals. All of the nine social scientists at top levels of the program were economists. The project designers chose a "package" program: introduction of high-yielding varieties of wheat in combination with fertilizer, pesticides, and technical assistance in farming methods. In addition, simple farm tools were redesigned.

In 1967, subsistence farmers in Ethiopia, a large majority of the rural population, were dominated by a rural elite that controlled large landholdings and the political system. Subsistence farmers also suffered from lack of roads and access to markets and educational facilities. No effective agricultural research or extension service existed anywhere in Ethiopia. The physical and economic infrastructure therefore was far inferior to that available in the other three project countries.

Project planners recognized that the conditions of land tenure would severely limit the benefits reaching subsistence farmers, but the government promised to carry out major land reform. Since the government was then controlled by a cohesive rural and urban elite, fulfillment of the promise proved politically impossible until the violent revolution of 1974.

CADU planners put their first emphasis on marketing, establishing thirty-three centers through which materials and information could be distributed, in an effort to assure small farmers a fair price for the expected increased production. These centers gradually expanded their activities, to selling high-yielding seeds, fertilizers, and other inputs recommended by CADU. Research in the development and adaptation of high-yielding wheat varieties and in determining the experimental basis for local recommendations for chemical fertilizers and pesticides was also emphasized.

CADU provided sixty-two extension agents who worked through "model farmers" in participating communities. (This idea came from exchange of visits with Comilla staff.) Each model farmer came from an area of approximately one hundred farm families, who collectively nominated five candidates. Final selection from among the five was made by the CADU staff. Only local full-time farmers who cultivated lands of average size for the community were eligible. Under the guidance of the extension agent, the model farmer used part of his holdings as a demonstration plot.

Up to 1967, small farmers had been unable to get any credit from banks and had been paying up to 400 percent on money borrowed from landlords and local moneylenders. CADU negotiated with the government bank a large loan at 10 percent interest and then administered its own credit program to farmers, charging them 12 percent. Along with this credit came supervision in the planning of the farming operation. This program increased the number of small farmers receiving credit from 868 in 1968–1969 to 14,146 in 1970–1971. The program was intended for farmers holding twenty-five hectares or less. In the first period, large farmers are able to secure 35 percent of the credit issued, but, in response to SIDA's urgings, the government agreed to restrictions against larger farmers so that by 1970–1971 only 2 percent of all loans went to farmers outside of the target population.

The great increase in wheat yields made possible by the new technology played into the hands of the large landowners. Since Ethiopia imposed no tariffs on imported tractors or fuel, large landowners found it profitable to dispossess their tenants and mechanize their operations. By 1974 five thousand tenant families or about thirty thousand total population had been pushed off the land (Cohen, 1975). Although tenants constituted 46 percent of farm families in 1968, they were only 12 percent by 1972.

Despite the catastrophic impact on tenants, we should not assume that the long-run impact of CADU was entirely negative. A greatly expanded corps of Ethiopian agriculture professionals was trained, who increasingly took over the direction and staffing of agricultural projects. Furthermore, lessons learned by SIDA and its Ethiopian counterparts resulted in the redesign of R&D strategy to produce a mini-package program (high-yielding seeds, credit, and extension), which began to have broad impacts on the countryside (Lele, 1975). Finally, the violent revolution of 1974 precipitated a drastic land reform program, eliminating the large landholders and distributing land to the peasant farmers. Although the disturbances following the revolution made it difficult to develop any stable program in Ethiopia, the drastic shift in land tenure appears to have provided a foundation upon which agricultural research and extension activities stimulated by CADU could have far-reaching positive effects upon the rural poor.

Puebla

The Puebla project in Mexico grew out of the work of CIMMYT, the international wheat and maize improvement center, whose program had grown out of a Rockefeller Foundation project begun in 1943. The contrast between CIMMYT's spectacular success in raising wheat yields and its failure to accomplish comparable gains in maize inspired some of the leaders of CIMMYT to undertake an intensive project with maize beginning in 1967 (CIMMYT, 1974).

Project planners recognized that contrasting experience with the two crops could be explained in large measure by differences in land tenure and access to irrigation. The most impressive gains in wheat yields had been achieved in northwestern Mexico on large landholdings effectively served by modern irrigation systems in which the government had invested millions of pesos. Then as now, however, maize was grown primarily by small farmers on rain-fed lands. The challenge to the Puebla project was to devise a method of improving the profitability of maize for thousands of small farmers without irrigation.

The first requirement for location of the project was an area where maize was the principal crop. The state of Puebla met this requirement. The project offices were located in the city of Puebla, less than a two-hour drive from CIMMYT and a similar distance from the graduate school of agriculture (Colegio de Postgraduados) at Chapingo, which is only a fifteen-minute drive from CIMMYT. The location had the further advantage that it was the native state of the president of Mexico and so won his enthusiastic endorsement. But CIMMYT did not work out any collaborative arrangements with the Instituto Nacional de Investigaciones Agrarias, the national government agricultural research institute (INIA), or with the extension service. Its only link to Mexican institutions was through the graduate school at Chapingo, which provided the field staff and implemented the design largely formulated by CIMMYT.

Although according to a project survey in 1967, 69 percent of the cultivated land in the area was devoted to maize, only 21 percent of total family income came from the sale of maize. Beans were the second most important crop, and all crops together accounted for only 30 percent of the income of Puebla farmers. Animals accounted for 28 percent, and off-farm income from labor and other activities for 40 percent. These figures indicated that the problem of raising family incomes involved more than maize production. Still, maize was a major focus of research and development for CIMMYT, and the planners were determined to improve maize yields.

> The action program of the Puebla Project was organized initially to include four major components: (a) varietal improvement of maize, (b) research to develop efficient recommendations on maize production practices, (c) assistance to farmers in proper use of new recommendations, and (d) coordination of the activities of the service agencies, the project team and the farmers. Another component—socio-economic evaluation—was added during the first year. [CIMMYT, 1974, p. ix]

Project planners soon discovered that they would have to abandon the objective of varietal improvement. Under the land and water conditions at Puebla, the native (or *criollo*) varieties did about as well as the improved varieties that CIMMYT was seeking to introduce.

Having decided to drop a line of activity which represented the greatest strength for an international research institution, project planners concentrated on the coordination of services, helping farmers to secure credit and fertilizer as well as providing instruction and recommendations for cultivation. CIMMYT reports an increase in average maize yield among project participants of 30 percent between 1969 and 1972. Although substantial, this increase is hardly spectacular compared to the much higher increases achieved elsewhere from high-yielding varieties of wheat or rice as part of new technological packages.

In assessing the adoption of project recommendations, CIMMYT used the number of farmers securing credit through the project for farm inputs, a rough measure at best. A large portion of those using Puebla credit to buy fertilizer did not actually apply the fertilizer according to project recommendations (Gladwin, 1979). Other researchers have observed the same tendency among small farmers in Central and South America. Where maize is grown chiefly for family consumption, farmers tend to buy fertilizer primarily for crops that are marketed. Even assuming that the CIMMYT measures of adoption were valid, results disappointed the planners. The numbers securing credit through Puebla increased rapidly in the first several years, then slowed down markedly. By 1973, CIMMYT estimated that only 26 percent of the land used for maize was covered by the Puebla program.

Disappointed by the declining rate of increase in the adoption of CIMMYT recommendations, Mexican professionals fortunately were not satisfied with the traditional answer that slow progress was the result of traditional farmers' resistance to change. They discovered that some farmers who had rejected the Puebla recommendations were doing better than some who joined the project. Out of their study of the more successful nonadopters sprang the discovery or rediscovery of peasant rationality.

An agronomist with the project described how the revelation came to him:

> I was talking with a farmer who interplanted squash and maize. I told him that was the wrong way to farm, that he should concentrate on maize. He argued with me, claiming that he would be better off with his own system. I couldn't believe him, but finally he proposed that we test his system against mine. He marked out two plots of equal size on his land. On one plot, he planted maize and also squash between the rows. On the other plot, he planted maize according to the recommendations I had given him. At harvest time we got together to evaluate the results. My plot did indeed look much neater than his, but when we came to measure the yield of maize, we found that there was no difference between the two plots. In his plot he had the additional yield of squash. That convinced me. [Alierso Caetano, personal communication]

To state simply a complex set of findings, the researchers discovered two agricultural production techniques among the more successful nonadopters: the rotation of corn and beans, and the interplanting of crops. In the rotational system, the first year the farmer planted corn supported by a heavy application of chicken manure. The second year, he planted beans without any fertilizer. For the third year, he reverted to corn but this time applied a small amount of chemical fertilizer—or else he shifted back to the first year's practice, with another generous application of chicken manure. In the interplanting system, the farmer planted beans between the rows of corn, either at the same time or several weeks later. The bean plants climbed the corn stalks (thereby getting better exposure to the sun), and, by approximately doubling the intensity of use of a small plot of land, the farmer was able to increase his returns very substantially.

The project's general coordinator (1970–1973) offered this insight:

In Mexico we had been mentally deformed by our professional education. Without realizing what was happening to us, in the classroom and in the laboratories we were learning that scientists knew all that had so far been learned about agriculture and that the small farmers did not know anything. Finally we had to realize that there was much we could learn from the small farmers. [Mauro Gomez, personal communication]

Interplanting of corn and beans was against the recommendations of the planners of the Puebla project. Struck by the obvious success of this violation of their rules, the researchers asked themselves: "Why have we been telling people they should not interplant corn and beans?" When they were unable to discover a solid rationale for the advice against interplanting, the researchers came upon the real source of that doctrine: "That is not the way the corn farmers do it in Iowa." In the U.S. corn belt, most of the work is done with tractors, which require space between rows. But the Puebla planners were not trying to introduce tractors. They simply followed U.S. planting practices without recognizing that the logic of the U.S.-style row spacing was keyed to the use of the tractor, which was not appropriate for small farmers with little capital and ample labor.

This revelation had a dramatic effect upon the direction of the Puebla project. Now the field staff began to learn from the farmers. By the fourth year of the project CIMMYT had discovered that "the studies of the maize-bean association demonstrated that net income from the association was approximately double that obtained with either maize or beans alone" (CIMMYT, 1974:28). By 1973 CIMMYT was able to offer farmers recommendations that included packages of production practices for the maize–pole bean association. In the same year CIMMYT finally adapted its organizational strategy to this combination, stimulating the formation of a union of progressive maize and bean farmers.

By this time, following a vigorous internal debate, CIMMYT had decided to withdraw from the Puebla project. Its own report provides this rationale:

CIMMYT decided in early 1972 to terminate its participation in the Puebla project at the end of 1973. The project had begun in 1967 as an experiment to learn how to rapidly increase maize production among small, low income farmers. As the project evolved, however, it became clear that the project's objectives would shift to more efficient strategies for increasing production, net income, and the general welfare of small farmers in rain fed areas. CIMMYT felt that its mandate was not broad enough to encompass all the activities that clearly should be incorporated

in so extensive an undertaking. This position was made known to the
Governor of Puebla and the Secretary of Agriculture, making clear CIM-
MYT's reasons for withdrawing support, as well as the conviction that
the project should continue. [P. 15]

In other words, CIMMYT came to recognize that its single-track strategy, concentrating on maize, was inadequate for dealing with the complex and interrelated problems of small farmers in Puebla. The lesson
was important. Staff writers seem not to have grasped another major
lesson indicated by their experience, however. The project report concluded with this statement: "Clearly, the job of adjusting and delivering adequate technology, as well as that of inducing farmers to use the
recommended technology, is very difficult and it is far from being
accomplished in the Puebla area" (p. 81). Of course, it is difficult to
get farmers to adopt "modern methods" when they can make twice as
much money with their own. The principal authors of the project report
clearly were still locked into the conventional definition of the problem
of "introducing change to small farmers": the problem was still seen
as how to transfer to those farmers a technology developed by the
professional experts. Fortunately, some of the leading Mexicans involved in the Puebla project broke out of that conventional definition
and used what they had learned in Puebla to organize a more creative
and efficient system of stimulating agricultural and rural development
among small farmers.

Caqueza

The Caqueza project in Colombia is notable for several reasons. (All
references in this section are to Zandstra, Swanberg, Zulberti, and
Nestel, 1979.)

Of all large research and development projects, Caqueza may well
be the most thoroughly documented, not only through agronomic and
plant science research but also through socioeconomic research. The
authors of the final project report are conscientious in describing
failures as well as successes. Furthermore, they concentrate particular
attention on the social and political processes involved in developing
the project.

Although the authors credit the Puebla project with being their principal inspiration, they make it clear that they did not simply follow the
Puebla design. They took advantage of the lessons that had emerged
when Puebla failed to reach its expected potential. As sometimes happens in the history of scientific development, even though the official
interpretations reported by CIMMYT are now subject to serious crit-

icism, that project had an enormously stimulating effect in leading researchers to explore a new strategy involving small farmers.

Caqueza planners started with the premise to which planners of Puebla ultimately came: that agricultural and rural development for small farmers must be based upon a broad and integrated approach. As in Puebla, maize was a major crop, but from the outset the planners expected to deal with a variety of crops and to experiment with intercropping. Providing credit for agriculture was an important goal, but the planners were also concerned with the health and educational needs of local people. In these nonagricultural fields, Caqueza was not to provide services directly but to help link small farmers with agencies having those responsibilities.

Caqueza planners sought to fit their project into the framework of agricultural R&D agencies in Colombia, working particularly with the Instituto Colombiano Agropecuario (ICA), Colombia's institute for agricultural research. Although accomplishment of local objectives would have been easier if it had remained independent, the initial and continued involvement of Caqueza with the national government helped to build the new strategies developed in the project into government programs.

Finally, Caqueza developed active collaboration with national faculties of agriculture, encouraging students and professors to carry out research of interest to Caqueza and providing some support for that research.

Caqueza was supported by the International Development Research Centre of Canada and by the government of Colombia. The project was placed administratively within ICA, which then defined its rural development goals as being to "generate and develop strategies to attack the restraints to social and economic development in specific geographic areas, characterized by the presence of subsistence farmers, through the incorporation of technologies which will rapidly increase the production of basic and traditional commodities in order to improve nutritional and income levels" (p. 31). These objectives appear to be narrowly based upon agriculture. Zandstra and his colleagues describe a much broader set of objectives for particular rural development projects such as Caqueza: "(1) To improve the standard of living, through improved community organization, housing, health, and education; (2) to increase the productivity of basic crops and animals; (3) to obtain efficient use of credit and market facilities; and (4) to encourage and obtain the integration of subsistence farmers' populations into associations and groups" (p. 31).

The project began in 1970 in an area containing between ten and fifteen thousand farm families, most of them small farmers, although a small minority held a large percentage of the land. Sixty-five percent of the farmers held under 5 hectares of land; their total holdings amounted to only 28.5 percent of the cultivated area. Twenty-two percent of the farmers operated farms between five and ten hectares, a total of 11 percent. At the other extreme, 4 percent of the farmers with holdings above thirty hectares accounted for 42 percent of the cultivated area. The cultivated areas ranged over three zones: below 1,800 meters in altitude, 1,800 to 2,300, and above 2,300.

Colombia had more people professionally trained in agriculture than most developing countries, but this educational advantage provided certain obstacles:

> During the first two years of the project a number of unforeseen problems became apparent. Paramount among these was the fact that most professional agriculturists working in Colombia had a training that was heavily biased toward large farm and plantation agriculture. Research was heavily oriented in this direction and was carried out almost entirely on large farms or on experimental stations that simulated large farm conditions. This situation was exacerbated by the fact that most graduate training took place in the United States and involved studying the problems of modern high input agriculture. As a result of this, knowledge of the local, complex, multiple-cropping, risk-aversion system practiced by the small farmer was extremely limited. [P. 10]

The project professionals devoted most of the first year to research on the actual farming practices and problems of the small farmers. In marked contrast to the original design of Puebla, the staff members did not assume at the outset that they knew what would benefit the small farmers and could simply adapt this known technology to local conditions. Their exploratory research produced important findings that shaped future activities. They found, for example, that 67 percent of the farmers already used credit for the purchase of fertilizer, but hardly any of them applied this fertilizer to corn or beans. They used it primarily for potatoes and also for vegetables and tomatoes.

The researchers recognized that this decision was economically rational, given the scarce resources of the farmers. Corn and beans were raised primarily for family consumption, and these crops would produce all or most of what was needed without fertilizer. On the other hand, fertilizer enormously increased the yields of potatoes, vegetables, and tomatoes—crops produced mainly for sale. Farmers recognized that although they could increase their yields of maize and beans substantially through application of fertilizer, the payoffs were much higher for the other crops. Furthermore, potatoes would barely produce in this area without substantial fertilizer. The farmers also recognized the great advantages of some intercropping practices. The researchers reported, for example, that since fertilizer is always used with potatoes, beans grown with potatoes yielded three times as much as those grown with corn.

The first year of research led to important conclusions regarding maize paralleling those in Puebla.

> Under the prevailing conditions of the area, the traditional corn varieties outperformed the improved varieties. The hybrids responded better to higher levels of fertilization (particularly nitrogen) than did the traditional varieties. However, they required better soil preparation and were more affected than were local varieties by low soil moisture and certain diseases. . . . New problems of post-harvest storage occurred because the hybrid grain was more susceptible to certain insects. [P. 63]

For farmers not accustomed to using fertilizer on their corn, the project leaders recognized that it would be a serious mistake to try to promote hybrid corn varieties that would pay off only with very substantial fertilization and better ecological conditions and better cultivation and storage practices than were likely to prevail in that area: "The old extension approach that considered communication of the new technology to farmers as the only activity required was being forgotten [by the end of the first year] and replaced by the idea that more had to be known about the farmers' present production system before anything could be done about changing it. But agronomic knowledge alone was not enough; socio-economic knowledge was required as well" (p. 64).

During the second year, Caqueza professionals found themselves grappling with major problems of linking the local farmers with the market for needed inputs and with the banking system.

> They soon discovered the administrative problems associated with credit programs and the difficulties small farmers had in obtaining the required amount of credit on time. The project staff were also confronted with the problem of insuring the supply of inputs specified in the loan programs. All too often these inputs were unavailable or did not arrive on time. . . . They also had to change their recommendations because the types of fertilizer and insecticide that they recommended were not available in the area. [P. 76]

The project also encountered a crisis in marketing, arising out of its very success in increasing yields. With commendable frankness they report: "The farmers adopted the project's recommendations for cabbage production, yields increased 3 to 5 fold but there was no market available to absorb the excess production. The farmers, angered, dumped sacks of their cabbages on the doorstep of the project's office and organized a demonstration" (pp. 87–88). Project staff people responded by contacting wholesale markets in Bogota, and two staff members spent three days as cabbage salesmen, trucking the produce to the Bogota market in project vehicles. The cabbage crisis led Caqueza to increase its emphasis on marketing.

In the early stages of Caqueza there was considerable friction between project leadership and the central administration of ICA. The intensity of the friction varied according to the personalities in the key positions of both organizations but was basically a structural problem. Project design called for a great deal more local autonomy for Caqueza than had been customary within ICA. Leaders of ICA were accustomed to directing the research from the central office, whereas Caqueza people insisted upon the need for flexible local planning.

The progress of the project was also severely hampered at times by sudden budget cuts imposed by the national government. There was a good deal of turnover among project professionals, but this should not be viewed as entirely negative because many of those who left the Caqueza project moved into important positions in ICA or in other agricultural agencies of the national government, thus helping to spread some of the new understandings arising from the project.

Events led project leaders to recognize the interrelation of a wide range of community interests and needs:

For example, in one "verada" in which the project was working, the community was extremely interested in obtaining electricity. Irrespective of the issues raised by the project staff, the farmers and their families always brought up the subject of electrification. Finally, the project personnel contacted ICEL (Columbian Institute for Electrification) and convinced them to start an electrification program in the area. As a result of this, the community became very receptive to suggestions from project staff.

Thus, another lesson was learned, namely that the farmers' priorities were not all production oriented. Often, health, education, and public utilities (water, electricity, roads, etc.) were considered more important. By disregarding the farmers' own priorities the project risks failing in its production activities.   [P. 90]

The project leaders recognized the importance of farmer organizations, but efforts in this direction were generally unsuccessful. A cooperative sponsored by Caqueza did not gain strong popular support or develop an effective administrative organization. The Caqueza prodevelopment committee was similarly unsuccessful, perhaps because it fell into the hands of traditional leadership. The town priest became head of the organization and led it into projects for embellishing the town plaza and planting flowers around the church, and the organization disintegrated. (Kenneth Swanberg, one of the co-authors of the Caqueza book, tells us that he observed in a 1981 visit to the project area that the farmer groups were functioning more effectively.)

Caqueza gave special attention to risk and the methods of measuring risk. This research led the staff to make an important distinction between risks of production and price, which could be compensated for in insurance schemes, and institutional risks, which do not lend themselves to such treatment. The production risk for a farmer trying out a new variety of seed could be covered by guaranteeing him compensation equal to the amount he lost through the innovation if it did not prove as profitable as his traditional method. Similarly, a price drop can be insured against by having the government guarantee a certain level of support prices. Although such measures can be too costly for a developing country, in principle production risks have an economic solution. There was no such solution to institutional risks, as the report states: "The latter were epitomized by the continuing problems of obtaining credit in time to use it for purchasing seed and fertilizer for the corn crops, and by the unsuccessful efforts to obtain insecticides and fertilizer packages in quantities small enough for use by the farmer with one hectare or less" (p. 195).

In other words, when considering the recommendations of the agricultural professional, the farmer needs to consider not only production and price risks but also those caused by institutional inefficiencies or failures that will prevent him from putting a package together at the right time to achieve a hoped-for result. This analysis led professionals to give attention to the organizational and interorganizational problems giving rise to institutional risks.

Project leaders discovered that the Caqueza farmers used sophisticated judgments in balancing risks with investments. Their decisions involved not simply a question of whether to use much or little fertilizer but on which crops to use it.

Whereas corn producers in the region employed limited cash inputs . . .

potato production is simply not possible without the use of substantial inputs.

Farmers appeared to concentrate the use of their cash on a relatively small part of their farms in production activities that provided a high return, be it at considerable risk. They appeared to keep the total risk to which they exposed their farm enterprise in balance by planting a large area to a low-input, low-risk crop such as corn.   [P. 204]

The Caqueza report provides evidence of yield increases in some of its applied crop research and demonstration projects, but the most important outcomes were the reorientation of the national agricultural research, development, and teaching programs. The report states that multiple cropping research, a negligible part of ICA's program in 1971, had by 1976 become a major activity. The project also served to strengthen the agronomic and socioeconomic research base necessary for work with small farmers.

The project played an essential role in the creation in the national university of a new master's degree program in rural development, in collaboration between the national university in Bogota and ICA. Up to this time, agricultural degree programs in Colombia had been specialized to particular academic disciplines. Now some of those who had participated in research with Caqueza worked together to establish a program that provided a major concentration upon socioeconomic theory and methods. Because the Caqueza project had continuing and profound effects upon the organization of agricultural research and development in Colombia and in the teaching and research in agriculture in the faculties of agriculture, it becomes impossible to trace the specific impacts. They extended far beyond the original site and resulted from actions taken by people and organizations far beyond the project area.

Conclusions

Several lessons may be drawn from these cases.

Dominance of the Socioeconomic Structure. All four projects were designed to aid small farmers. Whatever their differences in strategies and tactics, the resulting benefits must be explained in large measure as stemming from the differing socioeconomic structures in the project areas.

CADU operated in an area of gross inequality in landholdings, in the distribution of wealth, and in political power. Under these conditions, the more affluent and powerful people are bound to absorb a lion's share of the benefits. Although large numbers of small farmers undoubtedly benefited from the CADU materials, credit, recommendations, and marketing assistance, those who were most disadvantaged at the start of the project, the tenants, suffered disastrously because they were evicted from their holdings (Cohen, 1975).

In Comilla, the differences in wealth, power, and size of landholdings were not nearly so extreme as in CADU. Here again, however, the more affluent and politically powerful people benefited disproportionately once the project leaders' efforts to achieve equitable distribution slackened. The small farmers did benefit substantially, however, and there were no reports of tenant dispossession.

No record exists for Puebla regarding the equality or inequality of benefits according to size of landholdings. It is likely that the landholding pattern of the ejido area under study was sufficiently uniform that these problems did not assume importance. The report for Caqueza, an area with large differences in the size of landholdings, does not tell us whether large landowners benefited disproportionately, but it does document substantial benefits received by small farmers.

We can generalize from these cases that the more unequal the distribution of wealth and landholdings, the more inequitable will be the distribution of benefits even from projects designed to aid small farmers. The Comilla and Caqueza cases suggest, however, that effective local administration of an R&D project intended to benefit small farmers can reduce inequitable outcomes.

Orientation of Research. The projects differed markedly in the content of research and in the balance between research and demonstration. As originally designed, Puebla could hardly be regarded as a research project. The underlying assumption was that the designers already knew what was required to increase maize yields and that such an increase would benefit the farmers. To be sure, the project leaders recognized that a certain amount of adaptive research was needed to determine what varieties of maize would be best and what recommendations regarding inputs and methods of cultivation should be made. This adaptive research demonstrated that the maize varieties developed at CIMMYT offered no advantages over the native varieties under local conditions and that some changes in agronomic practices were promising.

Nevertheless, this discovery did not at first lead them toward greater interest in learning the elements of the indigenous farming system. They simply shifted their plans to abandon the introduction of new genetic materials and concentrated instead on making agronomic recommendations about spacing of seeds, amounts of fertilizer to be used, frequency of application, use of pesticides, and so on, supported by credit so the farmers could purchase the recommended inputs. Socioeconomic research was added to the plan in the early stages as an afterthought, and it assumed importance only as the Mexican leaders of the program recognized the value of studying the indigenous farming system.

CADU necessarily had to depend more heavily upon research in its early stages, for Ethiopia had no institutions comparable to CIMMYT or the Mexican national agricultural research program to provide improved genetic materials and the findings of agronomic research. CADU had to make a major investment in building an agricultural research program for Ethiopia. Beyond the plant sciences, research was concentrated on economics, and studies of social structure and political power were neglected. The economists could hardly fail to recognize the extreme inequality in the distribution of land and political influence, yet they seriously underestimated the effects this problem would have upon their efforts to help the rural poor.

Comilla represents a sharp contrast to Puebla in its initial emphasis upon socioeconomic research. The planners of Comilla did not start out to conduct a demonstration project because they recognized that they did not know enough about the area in which they were to intervene to have any confidence in what they would be able to demonstrate. The leaders of Comilla began to design interventions only after they had carried out extensive research, including many interviews and group discussions with local farmers and community leaders. Here also their strategy contrasted with that of the other two projects in placing primary emphasis upon the organizational aspects of rural and agricultural development. According to the diagnosis of the project leaders, the poor farmers would need to organize more effectively before they would be in a position to gain the full benefits of the high-yielding varieties, fertilizers, pesticides, and particularly improvement of the irrigation system.

Caqueza provided the broadest base of systematic research in the social and plant sciences. Because it was organized within the framework of ICA, the national agricultural research institution, it had an important impact on the course of the development of national agricultural research.

Production Emphasis. The projects differed also in the degree of production emphasis. In the beginning, Puebla was focused exclusively upon increasing maize yields. The planners intervened to facilitate the furnishing of credit and extension services, which they justified as being indispensable for achieving the goal of a maximum increase in maize yield.

CADU concentrated much of its efforts upon increasing wheat

yields, but that was not the focus of the initial activities. CADU leaders sought first to improve the marketing system so that the poor farmers would gain more benefits from the yields that were expected to follow. Comilla was the least oriented toward production, though the investment, through groups, in tubewells paid off quickly in increased yields. The planners hoped to increase crop yields, but they saw this not as a primary objective but as an outcome that would follow upon the building of a solid base of social organization supported by physical improvements.

Caqueza began with a substantial production emphasis but across a broader range of crops than the other projects. Caqueza gave more attention to marketing problems than the other projects except CADU and went further in analyzing and attempting to deal with the socioeconomic infrastructure affecting local farmers.

Farmer Participation in Decision Making. Here the contrast is between Puebla and CADU on the one hand and Comilla and Caqueza on the other. Puebla was designed as a unidirectional intervention in which small farmers were to be persuaded to accept the recommendations of the project. It was only as the leading Mexican members of the project recognized that adoption rates were slowing down and turned to studying the indigenous farming systems that they came to realize the importance of small farmer participation in decision making.

When CADU was launched, not only small farmers were excluded from participating in decision making; no Ethiopians of any status or discipline contributed toward shaping the project plan. Only after some months did Ethiopians trained by or through CADU begin to play significant roles in the project. Furthermore, until the revolution, there is no evidence that SIDA made any systematic efforts to involve small farmers in decision making. To be sure, SIDA professionals must have learned from talking with small farmers—directly or through interpreters—but such interaction is quite different from having plans specifically designed to stimulate small farmer initiative. It was only after the revolution, when small farmers became much more active in organizing and demanding assistance, that SIDA people came to adapt their plans to peasant initiatives.

In Comilla, peasant participation in discussion and decision making was secured from the very beginning. In fact, the needs, priorities, and sequencing of activities in the project design were developed out of active discussion between professionals and small farmers. Leaders of Caqueza similarly recognized the importance of peasant participation at the outset and made efforts to involve peasants in decision making and to stimulate the development of farmer organizations.

Specialization. Only Puebla was designed to promote a single crop; CADU and Comilla eventually concentrated on the major single crop in their areas. It is noteworthy that the project which began with the strongest monocultural emphasis evolved away from such specialization and into intercropping. In this regard Caqueza benefited from Puebla's experience and pursued a cropping systems strategy from the outset. 'None of the programs encompassed animals as part of the farming systems. Yet we have seen that 30 percent of the income of the Puebla farmers came from animals, and undoubtedly they contributed to family nutrition and income.

From Pilot Project to Regional or National Program. When a pilot project appears successful, political leaders naturally want to extend its benefits as rapidly as possible over as wide an area as possible. Leaders of the pilot project may caution against such a policy, but they are likely to be overruled by national policy makers. As seen in the case of Comilla, such rapid expansion of a project designed to aid poor farmers tends to result in the richer and more powerful appropriating the major share of the benefits.

In the case of CADU, the underdeveloped state of Ethiopian human and material resources imposed a healthy restraining influence against too rapid expansion. Furthermore, Ethiopia had the advantage of not needing to impose a new model for agricultural research and development upon an entrenched bureaucratic structure, for no such structure had previously existed.

Two developments favored an effective expansion program. Recognizing that it would be impossible to duplicate CADU throughout the country, SIDA and the government settled for a minimum package program. Then the 1974 revolution removed the socioeconomic and political barriers preventing the rural poor from fully enjoying the fruits of research and extension. Following land reform, the benefits received by the rural poor depended upon the ability of the new agricultural research and development organizations to provide them with useful information and economic and technical assistance. (As yet we know little about the performance of these organizations following the revolution.)

The further development of the Mexican program, growing out of the Puebla experience, will be discussed in Chapter 12.

In contrast to the other projects, whose relations with national programs were exceedingly weak, Caqueza began as a recognized part of ICA, the national agricultural research organization. This position presented many problems in project implementation, and yet it provided channels through which Caqueza was able to gain important influence in reshaping the national program.

Linking Research and Education. CADU and Comilla appeared to have no formal linkages with the national systems of university education. Puebla began with a base in the Colegio de Postgraduados at Chapingo as well as in CIMMYT. The university involvement seems to have been for the purpose of assisting in the scientific direction of the project, with only incidental attention to the opportunities the project could provide to enrich the education of students and professors.

Here Caqueza shows a marked contrast. In the early stages, project leaders recognized the mutual advantages in involving students and professors in the national universities and in faculties of agriculture in reseach on problems of interest in Caqueza. The policy of fostering research opportunities for university people through Caqueza had two important advantages for Colombia. A broader range of research was possible than project personnel could have carried out by themselves, and the involvement of students and professors in such studies provided some of the future leaders of Colombian agricultural education and administration with an understanding of the strategy of involving small farmers in agricultural research and development.

Need for Linking Organization. Leaders of all four projects came to recognize the great difficulties small farmers faced in coping with the forces and organizations in the socioeconomic environment beyond the village boundaries. To some extent, officials of each project sought to link small farmers with the outside world. Caqueza went farthest in this direction. The authors of the Caqueza report stress as one of their major conclusions the need to develop what they call a "buffer organization." The underlying idea is important, but the naming of the concept is unfortunate. The dictionary describes a buffer as an object that cushions the shock when two objects collide. What is clearly needed is not so much a defensive and protective unit but an organization that takes the initiative to help farmers work out more effective relations with the market and government agricultural agencies. We will examine such cases later.

Need for Interdisciplinary Research. The projects clearly demonstrate the need for interdisciplinary research that transcends the traditional barriers between the biological sciences and the social sciences. A project that is built upon too narrow a research base (as became especially evident with Puebla) is bound to encounter problems that can be resolved only through broadening that base.

Before we can apply the lessons learned from the examination of past experience so as to design a new organizational model for agricultural R&D, we need to understand the complex physical and biological factors that establish both limits and possibilities shaping the development of this new model. We turn to that topic in Part II.

Part II

Physical and Biological Bases
for Small Farm Development

The purpose of Part II is to lay out, in the simplest terms possible, the nature of the physical and biological elements that must be taken into account in research and development aimed at helping small farmers in developing countries. We begin with climate and ecology and follow with chapters on working with soils and water, cropping systems, farming systems including animals, and tools and machines. The first two chapters deal with the two major environmental conditions that determine what crop, animal, and management systems are appropriate to an agricultural region. Chapters 6 and 7 discuss cropping systems and farming systems which include animals used by small farmers and their relation to the ecologies of different agricultural regions. Chapter 8 discusses the choices available to small farmers in the use of power and machinery and their relation to farming systems.

Most of the discussion is limited to the tropics, between 23° north and south latitude in Asia, Africa, and the Americas, where the majority of the world's less developed regions and small farmers are found. The regions of our concern are, in the Americas, from Mexico to Peru, most of Africa, and, in Asia, from India and Thailand to Indonesia.

We have tried to present an analysis of experience which may help us to see how the manifold options for management of small land units may be sorted out rationally and how combinations of crops, animals, and management practices may be put together into farming systems that are most likely to satisfy the needs of small farmers in specific ecological sites.

In Part II we concentrate on basic environmental and management constraints and opportunities for small farmers. National and international policy makers have to think of national and international food production in relation to burgeoning populations and set goals for increasing agricultural land or intensifying land use, but we are concerned here only with the most important factors that limit the ability of small farmers to attain their goals. Later chapters deal briefly with some of the larger policy issues and the degree of coincidence between the goals of the small farmers and those of the society. Here we simply attempt to give some background for understanding the relationship between physical environments and the farming systems that are available to farmers.

4

Climate and Ecology

Effects of Climate and Ecology

Farmers' opportunities to put together successful farming systems are determined to a large extent by the climatic conditions in which their land is situated. The dominant components of climate are temperature, precipitation, and solar radiation. Their maxima, minima, and seasonal course determine what crops grow best and when they can be grown most successfully during the growing season. Climate also is a critical factor in the weathering of the geological parent materials from which soils are formed. Since the climate of a region may have changed over long periods of time, the effects of these changes on soil formation may have had profound influences on the development of existing soils—the degree to which they are eroded or have been formed from alluvial materials. Together with the soils that have been weathered from native geological materials and those that have arisen from ice or water flow, wind action, or volcanic deposits, climate determines the natural plant and animal ecologies from which agriculture evolves.

Thus the evolution of farming systems in an area is dependent directly and indirectly on the conditions of its climate. The farmer's ability to benefit from the climatic aspects that favor productivity of his crops and animals and to mitigate those aspects that may limit their productivity sets the limits of his success.

Within the tropical zone there is only a small seasonal variation in day length—less than two hours at the outer limits and none at the equator. There is a considerable range of average daily temperatures, largely dependent upon elevation, but little seasonal change attributable to day-length changes. Rainfall varies greatly under the influence both of seasonal wind patterns, such as the trade winds and the monsoons, and of regional and local geography. The *World Maps of Climatology* (Landsberg et al., 1965) identify five major tropical climates, based on the total rainfall and its annual pattern, with subdivisions of two of the zones (tropical humid and tropical dry). They are as follows:

55

(1). Tropical rainy climates with or without short interruptions of the rainy season (12 to 9½ humid months and no more than 2½ dry months); evergreen tropical rain forest.

(2) Tropical humid summer climates with 9½ to 7 months humid and 2½ to 5 months dry; evergreen forest and humid savanna.

(2a). Tropical humid winter climates with 9½ to 7 months humid and 2½ to 5 months dry; deciduous transition forest.

(3). Wet and dry tropical climates with 7 to 4½ humid and 5 to 7½ arid months; mixed deciduous and evergreen forest and dry savanna.

(4). Tropical dry climates with 4½ to 2 humid and 7½ to 10 arid summer months; tropical thorn and woody perennials, and savanna.

(4a). Tropical dry climates with humid months in winter; tropical semi-desert vegetation.

(5). Tropical semi-desert and desert climates with less than 2 humid and more than 10 arid months; tropical semi-desert and desert vegetation.

In general, agriculture in the absence of irrigation is most important in the first three climatic zones. The majority of the tropical wet and humid regions are found in the band between 10° north and 10° south latitude. Tropical wet and dry, dry, and semi-desert to desert regions are more common outside of the 10° band next to the equator. But there are exceptions to both of these generalizations.

Since vegetation is the basis for both plant and animal life, ecologists have identified plant and animal ecologies in terms of the dominant climatic zones that constrain them, and their terminology is frequently based on the undisturbed natural vegetation that occurs in the zones. Thus tropical rain forest is associated with the wet and humid tropical climates, mixed evergreen and deciduous forest with the transition from wet to dry climates, and savanna (grassland) with the wet-dry and dry climates.

Effective Precipitation

Because of the profound influence of temperature on relative humidity and thus on evapotranspiration[1] a given amount of rainfall is less effective in promoting plant growth when evapotranspiration is high than when it is low. Figure 4.1 depicts the effects of evapotranspiration on the potential growing period of plants in different tropical climates. PET stands for the ratio of precipitation to evapotranspiration. A PET of 1 indicates that precipitation is equal to evapotranspiration in a given

[1]Evapotranspiration is the loss of water by evaporation from plant and soil surfaces.

Figure 4.1. Four types of crop-growing periods and their relationship to rainfall patterns and potential evapotranspiration

1. Moderately long growing season

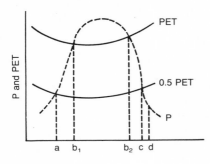

2. Short, predictable growing season

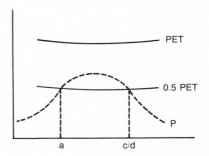

3. All-year-round humid, 12-month growing season

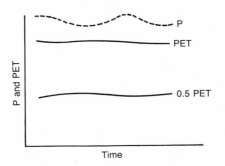

4. All-year-round dry, no predictable growing season

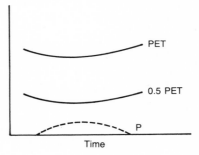

Source: FAO (1978: 38).

Key:

a	Beginning of rains and growing period
b_1 and b_2	Start and end of humid period, respectively
c	End of rains and rainy seasons
d	End of growing period
P	Precipitation
PET	Potential evapotranspiration

time period. It is estimated that a minimum PET of 0.5 is required to provide enough soil moisture for the beginning of sustained plant growth and that plant growth may continue for a time beyond the decline of the PET below 0.5 at the end of the growing season. The figure portrays hypothetical growing period curves at levels of 1 PET, 0.5 PET, and P (the actual precipitation) for the four principal climatic situations: (1) wet-dry climate with approximately eight months growing season; (2) dry climate with approximately five months growing season; (3) rainy and humid climate with a year-round growing season; and (4) semi-desert and desert climate with no well-defined growing season.

Although the annual and monthly average rainfall and evapotranspiration give a general picture of the ecological constraints for precipitation, it is also important to the ecologist and the farmer to understand the probabilities that surround these averages. In some areas within a climatic region there is far greater certainty that the average precipitation for the growing season will be approached in a given year than in other areas of the same region. As Hargreaves (1972) has suggested, a probability of 80 percent provides reasonable assurance that a given amount of effective rainfall will be obtained in four years out of five. The risk of crop failure or a serious drop in production in more than one year out of five may be acceptable to well-capitalized commercial farmers but not to the small subsistence farmers in the same area.

Crop Ecology

The production of agricultural crops in a region depends upon their adaptability to climatic ecology. One of the most thorough geographical studies of crop ecology has been undertaken by the FAO Agro-ecological Zones Project (FAO, 1978).

Using the available phenological information for eleven of the principal food crops[2] of the world, the project has undertaken to identify the geographical zones in which these crops may be expected to yield well enough to satisfy the needs of farmers at two levels of technology: low and high inputs of fertilizers, pesticides, machine power, and manual labor. Small farmers fall in the low-technology category and are found to a large extent in the more marginal soil and rainfall parts of

[2]Information on the time sequences of growth, flowering, and maturation of species and their cultivars as controlled by their genetic makeup. The crops are millet, sorghum, maize, wheat, soybeans, phaseolus beans, sweet potatoes, white potatoes, cassava, cotton, and rice.

the agroecological zones for the different crops. Thus, although the generalized maps showing zones ecologically satisfactory for the primary food, cereal, legume, and root crops may be helpful in planning regional agricultural development programs, they cannot determine the options of the small farmer within a zone or the best ways for him to put together a cropping system that provides for his subsistence and market needs. Nevertheless, they are useful in indicating the general climate limitations for these important food sources in particular locations, and the procedure developed for the project is one that should be of value for an understanding of other crops in relation to agricultural regions.

The implications of climate and ecology for small farmers will be further explored in the following chapters of Part II.

5

Working with Soils and Water

The greatest potential for increasing agricultural productivity of soils and increasing the amount of arable land under cultivation is in the tropics where most developing countries are located. This chapter presents a brief description of the soils and their potential within the different climatic subdivisions, classified according to Landsberg et al. (1969).

Soils of the Rainy Climates

These are regions of dense forest vegetation with 80 or more inches of annual rainfall well distributed during the year except for two or three months of low rainfall (less than 100 mm) in some areas. Average daytime temperatures are generally between 80 and 90° F. High humidity and small diurnal and seasonal temperature fluctuations characterize the climate. These regions lie mostly within 10° north and south of the equator. The Amazon basin in South America, Zaire, and southern coastal regions in Africa, parts of Indonesia, the Philippines, and other Southeast Asian countries are largely within this belt, which comprises about one-fourth of the total area in the tropics.

Most of the upland soils in the region are highly leached and weathered and consequently are of low fertility.[1] As used in this chapter and by most soil scientists, fertile soils are those with the capacity to supply nutrients to plants in adequate amounts and in proper balance for good crop production, with no toxic substances present to affect crops adversely. A fertile soil is not necessarily a productive soil unless there are an adequate moisture supply, good drainage, and no serious physical constraints such as a hardpan subsoil layer or steep topography. Some soil scientists and most laymen equate a fertile soil with a pro-

[1]These soils are classified largely as ultisols and oxisols in the U.S. soil taxonomy (Soil Survey Staff, 1975), and as ferralsols and aerosols in the FAO soil classifications (FAO/UNESCO, 1974).

ductive soil, but it is useful here to distinguish between the two concepts.

Though most soils of the humid climates are generally infertile, there is a substantial amount of relatively fertile alluvial soil[2] along the major rivers, encompassing some of the main rice-growing areas of the world, especially in Southeast Asia. Though representing only about 12 percent of the total area of the rainy climates, these soils supply a large percentage of the food produced in the region, primarily through intensively flooded rice (paddy) cultivation on small farms. Most of the population of this climactic zone is concentrated in these areas. Though most of the alluvial soils are relatively fertile, paddy rice yields are low, mainly because of inadequate water management. Poor drainage is a significant constraint in many areas and is related to poor soil physical properties and high water tables. High acidity and iron toxicity are also limiting factors in some areas. Additions of nitrogen and phosphorus and other nutrients increase yields substantially on alluvial soils, if other limiting factors are controlled. The highest rice yields in the world are on small farms on fertile alluvial soils in South Korea, where substantial amounts of fertilizer are used annually; fertilizers, pesticides, and other inputs are subsidized; adequate prices are paid to farmers to provide incentive; credit and marketing arrangements are favorable; and extension services are effective. All these factors, not just soils, contribute to the record high yields.

In some paddy areas two or three crops of rice can be grown on the same land in a single agricultural year. Skilled soil and crop management practices are necessary such as water control, appropriate fertilizer and pesticide applications, use of adapted varieties, appropriate timing of planting and harvesting, and grain-drying facilities. The long periods of cloud cover in much of the rainy region reduce solar radiation considerably, lengthening the period for the crop to mature and limiting the annual potential to one or two rice crops. Short-season crops such as vegetables can be grown during the few months of lower rainfall between rice crops.

Thus in many of the developing countries in the rainy tropics, substantially increased rice production on the alluvial soils is feasible with modest changes in the farming system: use of improved varieties and cropping patterns, water control, and application of fertilizers and pesticides. For these changes to occur, effective government services such as education, technical guidance, credit, guaranteed prices, and other incentives must be provided.

[2]These are classified as entisols in the U.S. Soil Taxonomy (Soil Survey Staff, 1975) and fluvisols in the FAO soil classifications (FAO/UNESCO, 1974).

In contrast to the relatively fertile alluvial soils, highly weathered infertile soils of the uplands comprise the largest area (about 75 percent) of the tropical regions having rainy climates (President's Science Advisory Panel, 1967). These soils pose serious constraints to the development and maintenance of permanent agricultural systems other than tree crops such as rubber, cacao, and oil palm. The predominant annual crop system on these upland soils is shifting cultivation (swidden or slash-and-burn) agriculture, the practice of cutting trees and other perennial plants, usually by hand, burning most of the vegetation during the period of lower rainfall, and planting mainly food crops for one to three years until yields decrease substantially because of decreased soil fertility and increased weed growth. The land is then allowed to revert to natural vegetation for ten to twenty years, depending on soil and other conditions, before being cleared and cropped again. In contrast to intensive rice cultivation on alluvial soils, shifting cultivation is an extensive system, which requires ten to twenty hectares or more per person. It is widespread over large areas of the tropics.

Within the rainy climates, there are some small pockets of upland soils which have relatively good nutrient status. These soils, derived from basic rocks or relatively recent volcanic ash on level or moderately sloping land, have a potential for a viable annual crop production system requiring modest inputs. An example is the Altamira area along the trans-Amazonian highway in Brazil. With shifting cultivation on these soils, the cropping period may be extended and the fallow period shortened. Tree crops such as rubber and cacao produce moderately well with little or no fertilizer.

Soils of the Humid Seasonal Climates

Under natural conditions these regions support predominantly deciduous forests in which most species lose their foliage during the dry season, which lasts from three to five months. Substantial areas of the region are no longer forested because vegetation has been burned over many years during the dry season, resulting in failure of the forests to regenerate. Man-made savanna becomes the permanent vegetation, with tall, coarse grasses predominating. This is the largest climatic region of the tropics, comprising about 30 percent of the total area (President's Science Advisory Panel, 1967). There is sufficient rainfall during the rainy season (seldom less than 100 mm per month) to satisfy the needs of a large range of rain-fed crops.

As in the region of the rainy climates, the highly weathered, infertile soils (oxisols and ultisols) are by far the largest single group, covering about three-fourths of the region in sparsely populated areas (President's Science Advisory Panel, 1967). Most of these soils are under shifting cultivation. Where the forested areas have been replaced by man-made savanna, erosion may be a serious problem on sloping lands. During the several months of the dry season, vegetation is sparse, so that when heavy rains begin there is danger of accelerated erosion and loss of topsoil, especailly where infiltration is retarded by a relatively impermeable subsoil layer and the topsoil has a sandy or loamy texture.

There are also large areas of highly weathered, infertile soils (oxisols and ultisols), which are quite permeable and where erosion is minimal, in Zaire, for example. These soils can be cultivated continuously on moderate slopes without much soil loss, provided good management practices are used. Land clearing is perferably done by hand, and the vegetation burned during the dry season supplies nutrients to the subsequent crop. Mechanized land clearing is more costly and compacts the soil, which reduces infiltration of rainfall, making erosion more likely. Partial removal of the topsoil by land-clearing machinery and deposition in windrows or low spots are other disadvantages. And, with mechanized land clearing, burning of the vegetation to supply nutrients is usually not done.

Despite the disadvantages, rapid conversion to agricultural use of potentially productive soils in the forested areas requires land clearing by mechanical means. Hand clearing is slow and difficult and severely limits the amount of land a farm family can cultivate. Improved methods of mechanical land clearing have been developed in which the trees are sheared at ground level and the logs are brush piled and burned during the dry weather. Farmer cooperatives with sufficient capital as well as commercial plantations have used mechanical methods for land clearing successfully.

For successful continuous cultivation after land clearing, fertilizer or manures are required to maintain productivity on the highly weathered infertile soils. In the forested areas, when the vegetation is burned, the ash supplies some nutrients, and there is usually enough nitrogen for a modest crop for several years. In man-made savanna areas, where the forests have not been regenerated because of shifting cultivation, the application of nitrogen is necessary either as chemical fertilizers, in animal manures, or legume (nitrogen-fixing) crops in rotation. Where adequate fertilizers and lime have been applied and weed control maintained, continuous cultivation has been successful. Where crop residues and straw are available and have been used for mulching in lieu of fertilizers, productivity can be maintained. In addition to supplying nutrients, mulching conserves soil moisture, keeps the soil temperature lower, and helps to prevent erosion. The use of mulch along with minimum tillage has been demonstrated as a viable practice for continuous cropping.

As in the rainy climate regions, relatively fertile alluvial soils (entisols) comprise an important area (about 8 percent) in the humid seasonal climate zone. The main rice-producing areas in Asia are in this zone, as is most of the population. The potential for substantially increased rice production on these soils is similar to that on the alluvial soils in the rainy climates, and so are the main soil-related constraints. In addition, the longer dry period is advantageous for the harvesting, threshing, and drying of the crop. Also the longer daily period of solar radiation in most areas may provide a greater potential for high productivity with improved management practices, despite the shorter growing season.

Moderately weathered soils and soils of relatively high nutrient status occur in somewhat larger areas than in the rainy climates. As in the rainy climates, these soils under shifting cultivation can be maintained with a longer cropping and shorter fallow cycle than the highly weathered soils and offer a greater potential for continuous cultivation. They comprise only about 5 percent of the total humid seasonal climate zone.

Soils of the Wet-Dry Climates

These are regions ranging from open woodland to savanna or grassland natural vegetaion. Under natural conditions they comprise about 20 percent of the total area in the tropics. Large areas occur in south Asia, Africa, and central Brazil. There is no rain for five to seven months, but during the rainy season there is enough rainfall (40 to 60 inches) for at least one crop. Rainfall can be erratic, however, and long enough periods of drought may occur during critical growth periods in the rainy season to limit crop production severely in some years. Practices to conserve soil moisture or use of supplementary irrigation are often necessary for optimal production. Because there is a great amount of sunshine during the dry season, crop production under irrigation is successful on suitable soils. Land clearing is much less difficult than in the forested areas, which is a distinct advantage in the initial period of agricultural development.

Although extensive areas of highly weathered, infertile soils (oxisols and ultisols) occur in the region, for example in the central plateau of Brazil, moderately weathered and nutrient-rich soils occupy significant areas. These more fertile soils can maintain a subsistence level of crop production without fallow periods for much longer time than can the highly weathered soils. They are commonly low in nitrogen and sometimes phosphorus, so that manures or fertilizers are required for production beyond a subsistence level. Where there are no severe constraints such as steep slopes, very sandy soils (psammento and inbordis of entisols), poor drainage, salinity, and other problems, these soils can be made highly productive with modest inputs. Most of them are probably already under cultivation or in pasture, but the productivity is generally low.

Because of the long dry season and insufficient vegetative ground cover, sandy soils in this climatic region are subject to considerable wind erosion. Management practices to conserve soil moisture and maintain a ground cover can minimize soil loss. There is surface crusting on some soils during the dry season, and with the onset of rains there is danger of water runoff and serious erosion on sloping lands with little or no ground cover.

An important group of nutrient-rich soils known as vertisols or dark tropical clays occurs widely in the wet-dry tropics on level or slightly sloping land but not in extensive areas except in western India, Sudan, and Australia. These are fertile grassland soils and can be highly productive if nitrogen supplements are provided. The main limitation is the difficulty in cultivation because of their high clay content, which requires high power, either animal or mechanical. For this reason most of these soils are not cultivated under traditional agricultural systems and have remained largely in natural pasture. Where appropriate power and cultivation machinery are available, crop production has been successful. Cotton production in the Gezira scheme in the Sudan is a good example. As mechanical power becomes more available, more and more of these soils are being brought under continuous cultivation, both in large tracts and in small farm holdings. Millions of hectares of these soils throughout the tropics are still undeveloped, and they represent a likely potential for agricultural development.

Large areas of highly weathered, infertile soils are also found in the wet-dry tropics but are not as extensive as in the rainy and humid seasonal climates. Hundreds of millions of acres occur in the central plateau of Brazil (the Campo Cerrado) and in eastern Colombia (the Llanos). They grow native pasture of low productivity and are largely used for grazing cattle. Because so much of the land is level to gently sloping and has excellent physical properties, there is a large potential for development. Recent research has demonstrated that with good soil and crop management and fertilization, these soils are highly productive. Rapid development of the region is now under way, with both continuous cropping and pasture systems. Ground water resources are available for irrigation during the dry season in a significant part of the region, so that two crops a year can easily be grown.

Though alluvial soils are less extensive than in the higher rainfall zones, there are important rice-producing areas on these soils in wet-dry zones. Where irrigation is available, two or three crops of rice can be produced during the year. The soil constraints for rice production are similar to those in the other climatic zones.

Soils of the Dry Climates

These are regions having an extremely erratic rainy season of only two to four or five months. About 15 percent of the land area of the tropics is in this climatic region (President's Science Advisory Panel, 1967). Savanna grasses and thorn bush predominate in the native vegetation. The most extensive areas are the Sahelian zone in Africa and large parts of India and Pakistan. Rainfall is unpredictable, varying from five to 30 inches a year, which makes cropping hazardous. Only crops with minimum moisture requirements such as pearl millet are grown. Extensive grazing of native pasture species is common. Very low rainfall may occur several years in succession, resulting in serious crop losses. The unsuccessful British groundnut scheme in East Africa in the early 1950s failed largely because of several years of much below average rainfall (Wood, 1950).

Except for a generally low level of nitrogen, the soils are well supplied with nutrients, though they are not always in good balance. Salinity is a problem in some areas. Where water is available for irrigation, the soils can be highly productive. Without irrigation the unreliable rainfall can barely support subsistence living over long periods of time. Because of the long, hot, dry season and extensive areas of sandy soils without much vegetation, formation of surface crusts is common, making cultivation with hand tools difficult until the rainy season arrives to soften the soil. Planting is delayed and if the rainy season is short, the crop may not mature.

On land where soils were formed in geologic periods when rainfall was abundant, there are highly weathered, infertile soils, representing about 5 percent of the total.

There is a significant proportion (about 10 percent) of dark tropical clay soils in this zone. Though potentially highly productive, very little is under cultivation because the power requirement is lacking.

Water

Outside of the tropical rainy climates, where surplus precipitation during most of the year is a problem, the management of water resources in the other climatic regions where there is a pronounced dry season is a critical factor in making the best use of the soils under cultivation and putting into production land not now being cultivated. With increasing populations in the developing countries and rising incomes in many of them, agricultural development has been extended to areas in the wet-and-dry and dry tropics where the rainfall patterns may be erratic, and extended droughts are common. Much of the rain occurs in high-intensity storms resulting in substantial runoff and increased soil erosion.

It is estimated that in 1974 there were more than 240 million hectares of irrigated land in the world, or about 15 percent of the cultivated area, producing about 40 percent of the world's agricultural production (Levine and Coward, 1979). Many developing countries have plans to expand the area under irrigation substantially and to rehabilitate present irrigation facilities. India, for example, has a long-term strategy to double the gross irrigated area from about 45 million hectares in 1977–1978 to 92 million by 1992–1993. The investment is huge; at present India is spending $1.6 billion on major irrigation projects in addition to some $600–700 million on medium or minor projects. Average costs are $1,000 to $2,000 per hectare for each irrigation development.

In the past, in response to rainfall uncertainty, individuals, local groups, and governments have constructed relatively simple irrigation systems that do not depend upon water storage. The primary objective of these systems has been to provide short-term drought relief. Water management has emphasized reliability of water delivery rather than maximum efficiency of use. There has been a tendency to strive toward maximum yield per unit area because water has had a relatively low value except during occasional drought periods. These systems and the farmers they serve are now under increasing pressure to make more effective use of the water resource to meet the demand for agricultural products. Toward this end, greater emphasis has been given in recent years to storage of water in order to lengthen the irrigation opportunity and use the water more efficiently.

Traditionally, crop production has been concentrated in the wet season with supplemental irrigation to provide water in periods of rainfall uncertainty. Irrigation systems were used in the dry season only in limited areas. But in recent years there has been increasing recognition of the advantages of growing crops during the dry season: higher solar radiation; decreased incidence of crop pests and diseases; use of available labor; better use of existing irrigation processing and production facilities during the entire year rather than only during the wet season. Pressures to increase cropping in an economically optimal way at the national level have resultant implications for managerial structures and for use of investment and other resources. There are also environmental implications since the shift to dry season production increases the pressure to develop reservoir storage. In Asia (not only in India but also in the Philippines, Thailand, and Malaysia) major investments are being made in large reservoirs and associated irrigation and drainage systems.

Water excess also may be a problem in irrigated areas, especially in the arid and semiarid regions. A classic example is the Indus basin of Pakistan, where overirrigation has raised the ground water levels so close to the surface that the high evaporation rates bring large volumes of salts to the surface, making it impossible to grow crops on large areas of land. Effective alternatives for solution of drainage problems are costly.

Choice of Appropriate Water Management Technology

There is a wide range of technological approaches to the irrigation of agricultural areas in developing countries of the tropics, and the choice of those appropriate to particular situations depends on both physical and human factors. Extremely intensive systems may be appropriate where water is very limited, irrigable land exceeds the water supply by a considerable margin, the economic demand for the agricultural products is much greater than the supply, and the required social and institutional infrastructure are in place. These conditions are met in Israel, and as a result, the irrigation systems are very intensive. Water is delivered to individual farmers or collective units through closed pipes and distributed through trickle emitters whose discharge rate is based on consideration of the crop, the soil, and the climatic factors. At the other extreme, where water is plentiful and drought conditions less prevalent, as in rice areas of the humid tropics, river water is conveyed through unlined earth channels to service areas, where unmeasured flow rates are delivered through uncontrolled openings in the channel banks and carried over the surface from paddy to paddy. Between these two extremes, there are many possible combinations of physical infrastructure, organizational form, and operational practice that may be appropriate to particular situations. The more intensive the irrigation system, the greater the importance of effective social controls. Chapter 15 discusses irrigation research and development as they apply to local organizations and national or state agencies that control the effectiveness of irrigation systems.

In many of the developing countries, emphasis on reservoirs and more intensive irrigation can be expected to increase. This trend inevitably will increase the emphasis on water efficiency. These systems are expected to be economically viable, and there is a concern to serve the largest area possible consistent with other project objectives. The intensity of the irrigation system has direct implications for the technology that is required; whether this technology emphasizes physical factors or human factors has major implications for the type and magnitude of resource and investments necessary and for the probabilities for success of the project.

Minimization of Undesirable Environmental Impacts

Large-scale development of water resources and their application to croplands produce significant, and sometimes undesirable, impacts on the physical environment. Three such impacts are disruption of the natural regime of river systems, spread of disease and other health problems, and deterioration of land.

The construction of large storage dams on major rivers almost always has significant impacts on the associated river system, only some of which may be understood and anticipated at the time of planning. Fish populations may be altered; aquatic plant life may be affected; patterns of periodic flooding (some of which may be beneficial) may be changed. Similarly, the human populations living below the dam site may have, and certainly populations previously located within the reservoir site will have, significant changes thrust upon them. Usually those in the immediate vicinity of the reservoir experience the greatest costs, through dislocation, and frequently receive the least benefits from the completed project.

Upstream reservoir construction frequently introduces other new activities to the upper watershed area such as commercial logging, the spread of upland farming, or increased residential settlements. These activities may result in undesirable land uses which contribute to the physical deterioration of local resources as well as to problems of reservoir siltation.

Increasingly, project plans give attention to human resettlement and attempt to provide compensation and new opportunities for those dislocated. Understanding of the secondary impacts of large-scale construction in upper watershed areas has improved, and attempts are often to monitor these impacts and to control adverse effects. The techniques for dealing with these problems are by no means fully developed.

The increase in use of irrigation, especially year-round, almost invariably is accompanied by an increase in the incidence of diseases that have waterborne phases. Malaria can increase significantly; but known combinations of sanitation, mosquito control programs, and medication can keep such adverse impacts of the increased prevalence of water to a relatively low level. Of much more serious concern is the problem of schistosomiasis (bilharzia), a parasitic disease characterized by a cycle that includes human waste entering the water supply, a waterborne snail as intermediate host, and subsequent water contact by humans. Depending upon the particular type, schistosomiasis can markedly shorten human life spans or be generally debilitating. In some parts of the tropical areas with a long irrigation history, it is an endemic problem. Year-round availability of water can dramatically increase the population of the essential host snail. When coupled with the relatively poor drinking water and waste disposal situation in these areas, a major increase in the disease is almost certain to occur.

In spite of much concern over many years, there still is no effective control of the disease. Chemical control of the snails is extremely expensive; biological control has not been found to be effective on a large scale; human medication has not been feasible on the scale necessary; snail habitat control is difficult and expensive. It is possible to visualize a program that deals in an integrated way with the range of factors that influence the incidence of the disease and might be effective in minimizing its increase, but past experience with comprehensive programs that deal with physical, social, and economic factors in an integrated way is not encouraging. A probable increase in waterborne diseases must be considered when decisions are made about irrigation development.

The health problems of people are mirrored in the health problems of land, especially in the irrigated arid and semiarid regions. A combination of waterlogging and salinity has taken millions of hectares out of production and threatens millions more. Here, again, the source of the problem is known, in general, but the solution in specific cases frequently is difficult to achieve. Technical solutions generally are very expensive and, unless coupled with changes in irrigation practice, usually are temporary. These changes are in part physical, but they also include changes in attitudes, human interaction, and agency-farmer interaction.

The growing emphasis, if not dependence, upon irrigation as an essential component of the food production systems of developing countries suggests that the issues considered here will become increasingly important. Appropriate answers are not obvious, but major efforts are being made to find them.

National Programs to Increase Agricultural Production in Developing Countries

The following paragraphs deal with potential increases in production that may be expected from intensification of land use and from increasing the area of land under cultivation and the relationship of these activities to small farmers.

Intensification of Crop Production on Land Already under Cultivation. As part of the response to the urgent need for increased food supplies for their rapidly increasing populations, developing countries have undertaken programs to increase production on land already under cultivation. Use of fertilizers increased from 24 million tons of plant nutrients in 1957 to 98 million tons in 1977. In the same period, large increases occurred in the use of improved seed, pesticides, and other inputs, and the area of land under irrigation is estimated to have increased by 100 million hectares (Drosdoff, ed., 1979).

Although substantial increases in food production resulted from these improved practices, the total production barely has kept up with the increase in population and in some cases has fallen behind (Wortman and Cummings, 1978). Average production per unit of land has remained low in comparison to more developed countries. For example, according to the 1979 FAO production yearbook, the 1974–1978 average rice yields per hectare in India, Bangladesh, the Philippines, and Thailand were about one-third those in Japan, United States, and South Korea. In France, United Kingdom, and Mexico wheat yields averaged two and three times those in developing countries. The United States, Taiwan, and USSR reported average maize yields more than twice as high as those in Mexico, Brazil, and Kenya; and the latter

Table 5.1. Approximate cultivated and uncultivated but potentially arable land by continent (in millions of hectares.)

Continent	Total land area	Cultivated	Uncultivated but potentially arable[a]
North America	2,400	270	200
South America	1,800	80	500
Africa	3,000	190	540
Europe	1,000	210	160
Asia	4,000	700	200
Australia and New Zealand	800	50	100
Total	13,000	1,500	1,700

Source: These data are approximations adapted from President's Science Advisory Panel (1967), Dudul (1978), and Buringh (1982). 1 hectare is approximately 2.5 acres.
[a]Potentially arable land includes soils considered to be cultivable and acceptably productive of crops adapted to the environment. Some soils will need irrigation, drainage, stone removal, clearing of trees, or other measures, the cost of which would not be excessive in relation to anticipated returns.

three countries have substantially higher yields than India, the Philippines, and Nigeria.

The answer to the question, "What is the potential for further increases in crop production by intensification of land use?" appears to depend in large part on the soils and climates and the management systems in the different countries. Most of the increases in production resulting from improved technology thus far have occurred in the larger farms and better farmland of the developing countries. National food production programs in most developing countries are giving more attention to the small farmers, whose traditional farming systems have been relatively less affected by modern technology.

Since small farmers produce a large proportion of food for internal consumption, this new emphasis is necessary for the attainment of national goals. But much of the land they occupy is on hilly terrain, subject to erosion, low in fertility, and not served by irrigation. In such situations, it is either impossible or economically not feasible to attain as high yields as are attained on the better lands occupied by larger farm units. On the other hand, within the local physical constraints, improved farming systems that increase production per unit of land and are within the means of small farmers can frequently be promoted through national programs that provide the access to inputs and profit incentives necessary to motivate small farmers.

To a large extent, the success of the national agricultural development programs for the future will depend on their ability to mobilize and integrate the technological inputs necessary to provide for far more intensive land use in these tropical areas. The following two chapters deal with cropping and farming systems as they affect the intensity of land use by individual farmers.

Potentially Arable Land Not under Cultivation. Most of the natural fertile soils in the less developed countries (LDCs) with favorable topography and adequate water supply are probably now under cultivation. Therefore, greater research efforts are being directed to ways to use potentially arable lands, which presently have limited use because of difficult accessiblity, climatic limitations, soil and water constraints, and other factors. Table 5.1 summarizes by continents the estimated area of cultivated and uncultivated but potentially arable land.

The greatest potential for increasing agricultural productivity of soils not under cultivation as well as those under cultivation is in the tropics. By far the largest area of potentially arable soils not under cultivation is in South America and Africa. In Asia about 80 percent of the potentially arable soils are already under cultivation, but there remains a substantial area on that continent and on the other continents that could be brought under cultivation.

How much of this potentially arable land area will contribute ultimately to the support of the growing world population remains to be seen. Using the FAO methodology of assessing agroclimatic suitability for rain-fed production of pearl millet, maize, and cassava in Africa (FAO/UNESCO, 1974), it has been estimated that over 2,500 million hectares of land are suitable to production of these three crops, although only 40 million hectares are actually used for them. Soil limitations are not considered in these studies, including land suitability studies, and they reduce the potential area for production by 60 to 80 percent from the above figures. Another land suitability criterion not considered in the FAO agroclimatic studies is the hazard of soil degradation resulting from normal erosion, salinization, acidification, compaction, and other factors. These are being considered in a third FAO study now in progress. These various FAO programs are global in character, and interpretation on a regional basis requires soil surveys in sufficient detail to define site-specific conditions, information that is available for less than 20 percent of the arable land in tropical Africa, South America, and Asia. It is thus necessary, more often than not, for land-use planners in national institutions to put together admittedly incomplete information on the agroclimatic conditions and crop and land suitabilities in the prospective new agricultural sites in order to arrive at meaningful evaluations of their physical conditions in relation to the specific objectives of the national agricultural development plans. As yet, this information is very spotty and has not been carefully put together in a form usable for global evaluations.

6

Cropping Systems

In planning agricultural development research, one needs to understand the relations among the plants grown and their short- and long-term effects on the productivity of the soil. In an undisturbed ecological situation of the humid tropics several tiers of perennials (trees, vines, and herbs) and various successions of annuals may occupy a site in the course of a season. The biomass produced may be measured in dry weight, carbohydrate, protein, and fat or in energy. Part of it may have direct economic or subsistence value, part may furnish feed for animals, and part may serve to conserve soil fertility or check loss of soil and water. A stable cropping system makes appropriate use of the different climatic and land surface features through cultivation of perennial and annual plants. In such a system, there may be conbinations of perennials in permanent plantings similar to the undistrubed climax vegetation (natural forest, for instance) or economically more valuable but equally conservative woody vegetation (shade-grown coffee, for instance) or perennial herbs (banana, plantain, sugarcane, for instance) rotated with intensively grown annuals. The cropping system, then, is pattern and management of crops within an ecological and social environment.

Figure 6.1 gives a conceptual diagram of a cropping system. The figure is deceptively simple because it does not detail the ecological opportunities and management constraints within which the farmer must make choices. These will be dealt with in some detail as the chapter unfolds. .

Cropping patterns and crop management practices comprise the cropping system for a farm. The crops appropriate for a farm site are determined by their adaptation to the climatic conditions (temperature, rainfall, and light) and the terrain and soil conditions (mainly slope, drainage, and fertility) together with the management practices that are economically feasible to enhance the productivity of the site (fertilizer, water management, pest control). Social and economic constraints operate to determine which of the crops ecologically adapted to a site may

Figure 6.1. A conceptual diagram of a cropping system in relation to its environmental constraints, inputs, and outputs

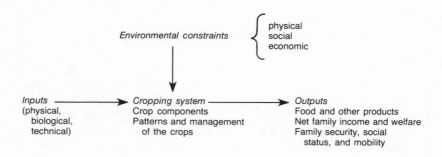

provide the best combinations for maximum net profit and/or subsistence in a stable system appropriate to the farmer located there.

Explanation of Cropping Terms

The intensity of land use at a site, of course, depends on the combined influence of the surrounding ecological, social, and economic conditions. Richard R. Harwood (1979) and Donald C. L. Kass (1978) have provided definitions of terms that are used in this chapter. If only one crop, either annual or perennial, is present and harvested in a year, the land is *monocropped* or in *monoculture*. The term most commonly used to indicate a land-use intensity greater than a single crop per year is *multiple cropping,* meaning that more than one crop is grown on a piece of land during an agricultural year. Within this definition, two or three crops planted and harvested alone, each after the other, are said to be planted in *succession* or *sequence.* If the two or more crops are planted at the same time or if perennials occupy land which is also cropped in annuals or other perennials, they are said to be *in association* or to be *intercropped.* If a crop is interplanted in another which is in process of growth, the second crop is said to be *relay planted* or *intercropped.* The term *rotation* indicates systematic changes in cropping over more than one agricultural year.

The intensity of land use in cropping systems may be measured in various ways. Perhaps the simplest index was suggested by J. H. K. Joosten (1962) and is used by Hans Ruthenberg (1971) to classify according to intensity of crop rotation, which includes years in which the land lies fallow. The symbol R is defined as the number of years of continuous crop cultivation multiplied by 100 and divided by the length of the cycle of land use (the sum of the number of years of arable crop farming plus the number of fallow years).

As long as fallow farming has an extensive character in which many fallow years follow a short period of cultivation R remains very small. If, for example eighteen fallow years succeed two years of cultivation, as may be the case in the rain forest, R amounts to 10. This extensive type of fallow farming is designated "shifting cultivation.". . . The larger R becomes, the higher is the percentage of the area cultivated annually in relation to total area available for arable farming and the more stationary the character of the farming becomes.   [Ruthenberg, 1971:3–4]

The most commonly used index of intensity of multiple cropping for a single cropping period is the land equivalent ratio (LER); it is derived by dividing the yield of each crop in a crop association by its yield in "pure" stand (Harwood, 1979). This index has provided valuable objective data for comparison of different cropping patterns within one or two planting periods.

Intensity of land use, in itself, is not necessarily an objective of a cropping system and is desirable only insofar as it contributes to the well-being of the farm family. More often than not, competition for light, water, or nutrients reduces the yield of the individual crops and labor is increased in an intercropping pattern, but in patterns that are adopted by small farmers the total cropping system results in increased productivity per land unit/time as measured in total yield or net economic gain.

Over a period of several crop years, the individual crop and intercropping patterns for different parcels of land in a small farm may be changed systematically in rotation to reduce problems with pests (including weeds) and erosion. Parts of the farm may be devoted to rotations of annual crops and other parts to permanent plantings of perennials, sometimes in association with annuals, to reduce erosion or provide pasture or firewood on land not suitable for cultivation.

Major Cultivation Systems

In the second edition of *Farming Systems in the Tropics,* Hans Ruthenberg (1976) has classified the different cultivation systems used by tropical farmers according to the type of rotation of land use:

(1) Fallow systems involving a long period during which the land is under a cover of trees and bushes, grass, or both, and a short period during which cultivated crops are grown. The shifting cultivation pat-

tern (also called swidden, slash-and-burn) is the most primitive of the fallow systems.

(2) Short-term rotation systems involving several years in which the land is devoted to grass and legumes used for livestock production, alternated with several years of arable cropping.

(3) Field systems on rain-fed land involving continuous use of the land for arable crops.

(4) Arable crop systems with irrigation.

(5) Systems with perennial crops involving long-term use of the land for tree or bush crops or for perennial field crops.

The fallow systems are least intensive, and the field systems are most intensive. The opportunities for intensive land use in the absence of irrigation are greater in humid and semihumid climates than in semiarid climates. The availability of irrigation water gives opportunity for intensification of land use through additional crops in the dry season.

Figure 6.2 gives a diagrammatic picture of the relationships of these basic management systems in categories according to intensity of land use.

The prevalence of one or more of these cropping systems in an area is determined partly by climatic and physical conditions (principally rainfall, temperature, terrain, and soils) and partly by social conditions (principally infrastructure, culture, population, and political policies). The systems are not necessarily fixed, even though there is a rationale for their existence. An understanding of the systems and of the reasons for them leads to innovations through application of modern technology that can bring improvements in productivity and social benefit.

Ecological Zoning of Crops within Farming Systems

In general, a crop's degree of adaptability to the principal ecological zones of the tropics is well known so that, with knowledge of the temperature and rainfall regimes, the crops best adapted climatically may be identified. Although less is known about the interaction of crops planted in association or in relay, the experience of farmers in many tropical areas with many crop associations may be drawn upon as a basis for initial determination of which combinations do well or poorly together in a given climate. Plant breeders in national and international insitutions are called upon to extend the ranges of climatic adaptability of crops in order to make it possible to grow them in areas to which their cultivars are only marginally adapted, so that new varieties selected for this purpose may be tested along with new varieties better adapted to the principal production zone.

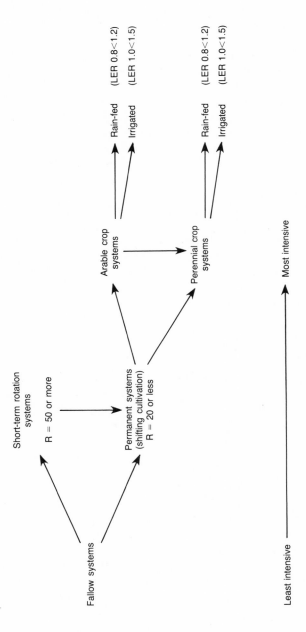

Figure 6.2. Relationships of basic management systems to intensity of land use

Short-term rotation systems

R = 50 or more

Permanent systems (shifting cultivation)
R = 20 or less

Fallow systems

Arable crop systems

Perennial crop systems

Rain-fed   (LER 0.8<1.2)

Irrigated   (LER 1.0<1.5)

Rain-fed   (LER 0.8<1.2)

Irrigated   (LER 1.0<1.5)

Least intensive   Most intensive

Source: Adapted broadly from Ruthenberg (1976: p. 16)

The sociopolitical impetus to extend the ranges of climatic adaptability of crops results from two somewhat different factors. One is the presence of a profitable market. If, for instance, the demand for white potatoes in Managua, Nicaragua, exceeds the national supply and the import price is high, development of a potato industry to supply the national market may be included as a national planning goal. The other factor is the food habits of the common people. Corn and beans are the preferred subsistence food staples of a large part of the population in Latin America. Rice is the essential subsistence food of most of Asia and is of great significance in tropical Latin America and Africa. This means that the rural populations grow these food crops not only in the local bioclimatic zones to which they are best adapted but also in adjoining areas where they are not as well adapted. Under these two motivations, breeders work against genetic odds to extend the productivity of corn in the highlands and lowlands, rice in dry upland areas and in areas subject to flooding, beans in high rainfall areas, and white potatoes in warm tropical areas, where the productivity of existing cultivars may be low. The satisfaction of these nonecological requirements may be difficult if not impossible; thus knowledge about adaptability of existing cultivars is essential in the selection of crops to be grown conservatively within a given bioclimatic zone.

Associations of Cropping Systems in an Ecological Zone

The categories of cropping systems described briefly above are found in various associations with each other and with forest or grazing systems, both within regions and within farms or plantations.

For instance, four permanent agriculture systems are short-term rotation, permanent arable crop cultivation on rain-fed land, systems of arable irrigated farming, and systems with perennial crops. All four of these systems may be found together in different combinations. In central and southwestern Costa Rica, small farmers in the Tucurrique zone (elevation 700 m, annual rainfall 3 m, no pronounced dry season) grow perennial tree crops (mainly pejibaye palm, citrus, and coffee), warm temperature vegetables for the Cartago market, and corn, beans, and cassava for home consumption. Some specialize in perennials and subsistence crops, others in vegetables and subsistence crops. Small farmers in the San Juan Sur zone (elevation 1,000 m, annual rainfall 3 m, no pronounced dry season) produce perennial crops (coffee and sugarcane) and subsistence crops. Small farmers in the Alejuela zone of the central plateau (elevation 1,500 m, annual rainfall 2 m, 4 months dry season, supplementary irrigation) grow vegetables for the San Jose market on their irrigated land and subsistence crops on their rain-fed land. In the same zone large farmers, who do not have irrigation, grow sugarcane, coffee, and corn as monocrops. On the western skirts of volcanoes Irazu and Turrialba at elevations of 1,800–2,000 m, with a 3-month dry season, there are examples of short-term rotation farming in which the improved grasslands are grazed by dairy cattle and the arable lands are used to grow cool-temperature vegetables (such as potato, onion, cabbage) and corn. The combinations among these four systems that are encountered in this small area of Costa Rica depend on rainfall pattern, availability of irrigation, mean monthly temperature, terrain, soil fertility, and local infrastructure. Since the population pressure in the area is not great and the soil structure is very stable, permanent cultivation on rain-fed land does not result in serious erosion. In other areas of the country where there are higher population pressures or the soil is less stable, the permanent cultivation practices used have caused very serious soil and water losses, which in some cases appear to have destroyed or irreversibly damaged the land resources within two decades following the clearing of the forest.

Cropping Systems Identified by Dominant Crops

It is usually possible to identify cropping systems within a climatic zone according to dominant annual and perennial crops. Table 6.1 classifies the principal cereals, tubers and roots, and seed legumes and the principal herbaceous and arboreal perennials found in small farm cropping systems of tropical Asia, Africa, and the Americas. The Life Zone concept of L. R. Holdridge (1967) is used to divide the major tropical climates into six groups which have different temperature and rainfall regimes. The dominant crop within the Life Zone depends on custom and market as well as on microclimatic conditions within a site.

The combinations among these dominant crops and others, including nonfood export crops and vegetables for local markets, in any agricultural site determine its cropping system. They may include two or three monocrops planted in succession, one or more of them relay-planted, or two or more planted simultaneously as intercrops. Trees or shrubs may be combined with each other in two or three "tiered" plantings. Together they comprise the cropping patterns for the agricultural year.

Important Cropping Systems in Southeast Asia, West Africa, and Central America

Examples of important cropping systems in Southeast Asia, West Africa, and Central America, outlined below, illustrate the relation-

Table 6.1. Cropping systems in different life zones, identified by dominant crops

Life zone	Dominant annual crops				Dominant perennial crops	
	Cereals	Tubers and roots	Seed legumes	Herbaceous	Arboreal	
Tropical rain forest 0–500 m altitude no defined dry season 1,500–3,000+ mm rain	paddy rice	dashien taro	cowpea string bean	banana, plaintain	rubber, oil palm, cacao, coconut, trees for forage, firewood, and lumber	
Subtropical humid forest (Premontane) 500–1,000 m altitude short dry season 1,500–2,000 mm rain	paddy and upland rice corn	cassava sweet potato	common bean string bean peanut	sugar cane banana, plaintain pasture grasses legumes	avocado, mango, citrus, cacao, oil palm, coffee, trees for forage, firewood, and lumber	
Subtropical dry forest (Premontane) 500–1,000 m altitude long dry season 1,000–1,500 mm rain	corn sorghum	cassava sweet potato	common bean peanut soybean	sugarcane pasture grasses and legumes	avocado, mango, citrus, coffee, trees for forage and firewood	
Temperate moist forest (Low montane) 1,000–2,000 m altitude long dry season 1,500+ mm rain	corn millet wheat barley	white potato	common bean broad bean chick pea	range and pasture grasses and legumes	citrus, mango, deciduous fruits, mixed forest trees	
Temperate dry forest (Low montane) semiarid 500–1,000 mm rain	sorghum millet wheat barley	white potato	broad bean (sweet) pea	range and pasture grasses and legumes	deciduous fruits, mixed forest trees	
Cool temperate dry to moist forest (Montane) 2,000–3,000+ m altitude long dry season 500–1,000 mm rain	wheat barley	white potato	broad bean	high-altitude range grasses	evergreen forest trees	

Classified according to Holdridge (1967).

ships between climate, crop combinations, and cropping patterns. Many variations are played on these themes.

Southeast Asia. Two examples illustrate the possibilities.

(1) Nonirrigated rice as the dominant crop.

Harwood (1979) has classified the rice-growing areas of Asia into five categories, according to the number of months during which the mean rainfall may be expected to be above 200 mm. Each category corresponds to a specific rice-growing potential.

Category I has less than three months with 200 mm of rain and is risky for production of a single rice crop during the rainy period.

Category II, with from three to five months having 200 mm or more of rain, includes the prime areas for growing a single crop per year of transplanted rice.

Category III, with five to seven months of rainfall above 200 mm per month, can provide for two crops of early-maturing rice.

Category IV, with from seven to nine months of 200 mm rainfall, can grow two crops of transplanted rice.

Category V is made up of areas with more than nine months with 200 mm of rain and can support continuous rice production (three crops).

Thinking of cropping patterns appropriate to the Central Luzon region of the Philippine Islands, which has more than 200 mm of rainfall during four months of the year, Harwood has presented (Figure 6.3) the hypothetical rainfall and standing water pattern, in relation to a crop of paddy rice preceded by green maize and followed by grain legumes (using bean or cowpeas) or vine crops.

(2) Irrigated paddy rice as the dominant crop.

Some of the most intensive cropping patterns that have been developed, with rice as the dominant crop are found in central and southern Taiwan. Chien-pan Cheng (1975) has described several patterns built around two rice crops with one, two, or three dryland crops planted in the time intervals between the rice crops. The general schedules for field operations of the two rice crops are as follows:

Region	Crop	Transplanting	Harvest
Central	1st	February–March	June–July
	2d	July–August	October–November
Southern	1st	January–February	May–June
	2d	June–July	September–October

Figure 6.3. Possible crop patterns in lowland rice areas having three to five months of good rainfall

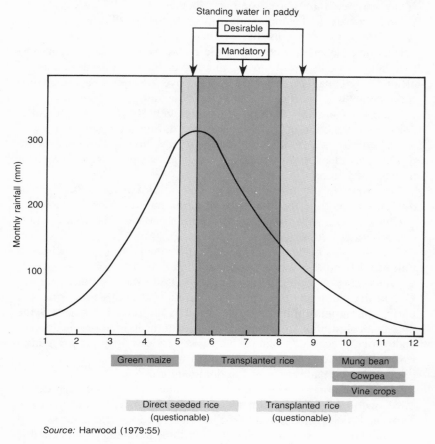

Source: Harwood (1979:55)

The leading cropping patterns involving two rice crops are the following:

1. Rice–Rice–Sweet potato in sequence
2. Rice–Rice–Soybean in sequence
3. Rice–Rice–Corn in sequence
4. Rice–Rice–Tobacco in sequence
5. Rice–Rice–Vegetables in sequence (with 90–100 day maturation period such as white potato and tomato)
6. Rice–Vegetables–Rice (Early maturing rice varieties are used for both rice crops. Muskmelon, pickling cucumber, or watermelon are relay-planted in the first rice crop about 25 days before it is harvested)

7. Rice–Jute–Rice (Jute seedlings about 20 days old are relay-planted in the drained paddy about 20 days before rice harvest)

West Africa. Tropical West Africa covers a wide range of climates. Mean annual rainfall varies from more than 1,500 mm (tropical rain forest) to less than 250 mm (desert and semi-desert). The zones of intermediate rainfall may have long dry seasons extending over five months or more or shorter ones extending three months or less. Elevations range from less than 300 m to more than 3,000 m. As a result, the natural vegetation of the agricultural areas includes rain forest, savanna, and montane vegetation appropriate to the specific climate.

The dominant crops of the different climatic regions, of course, differ in accordance with their climatic tolerances and limitations (Okigbo and Greenland, 1976; Okigbo, 1978). In the driest areas millet is dominant. In the wettest, plantain and root crops are dominant. In between these extremes going from wetter to drier areas, one finds rice, corn, and sorghum, respectively, as the dominant crops.

Some of the most complex intercropping systems encountered anywhere are found in the rain forest of tropical West Africa, where staple food crops, vegetables, and perennial fruit trees are intercropped. In the southeastern states of Nigeria the dominant food crops are yams, cassava, cocoyam, banana, and plantain, as well as some maize. These crops may be intercropped with groundnut, okra, pumpkin, melon, or leaf vegetables. In much of southeastern and midwestern Nigeria, oil palms produce important cash crops and may be grown in monoculture or mixed plantings of annual crops. In the midwestern states rubber and cacao may be found in association with annual crops.

Central America. As in tropical West Africa, Central America covers a wide range of climates. Mean annual rainfall ranges from more than 3,000 mm in some coastal areas to less than 500 mm at some intermediate to high elevations in the central uplands of some countries. The length of the dry season varies greatly both at low coastal elevations and in the interior. Thus in the Pacific region of Costa Rica which borders Panama, there is no well-defined dry season, whereas 500 km to the north the Pacific area close to Nicaragua has a dry season of four months or longer. At intermediate elevations on the Pacific side of the Isthmus of Panama the dry season is generally longer (three to four months) than on the Atlantic (Caribbean) side (two to three months).

The dominant crops for the wet lowlands having poorly defined dry seasons are upland rice, cassava, banana, and plantain. Cacao and oil palm are the dominant woody perennial crops. Corn and the common bean are grown in those areas but are handicapped by moist weather at harvest and are planted so that they will mature in the driest months of the year. At intermediate levels up to 1,000 feet, corn, common bean, and cassava dominate the annual food crops, sugarcane dominates the herbaceous perennial crops, and coffee, the woody perennial crops. At higher elevations up to 2,500 m wheat, white potato, and other cool-temperature vegetables, citrus, and warm-temperature-zone fruit trees become important, in association with dairy and beef production. Where the dry season is long (four to five months) sorghum may complement corn or may be the dominant cereal. At higher elevations, up to 3,500 m, wheat either complements corn or becomes the dominant cereal, white potato is the dominant tuber, and European broad bean the dominant grain legume. At these higher levels, temperate-zone fruit trees become more important.

Since maize is by far the most important cereal of the region, it furnishes a basis for many intercropping combinations and patterns found in small farms. Some of the more important ones, and trials for improvement, have been discussed in the final report of the Centro Agronómico Tropical de Investigación y Enseñanza (CATIE) on the first phase of its Small Farmer Cropping Systems Program (1979). They are discussed below.

Maize and Common Beans Intercropped. In this case either bush bean or climbing bean may be used, and the planting arrangements and spacing may vary. Both types of bean are planted at about the same time as the corn. If bush beans are used they are usually planted between the rows or hills of corn. If indeterminant climbing beans are used they may be planted in the rows or hills of corn and are supported by the corn stalks. Following maturation of the ears, the stalks may be broken above the ears, which are bent toward the ground to aid in the drying of the grain. C. A. Francis has reported on studies of timing and spacing of maize–bush bean and maize–climbing bean combinations at the Centro Internacional de Agricultura Tropical (CIAT) in Colombia (1978). CATIE (1979), working under conditions of lower fertility in Nicaragua and Costa Rica, has reported on the importance of soil fertilization in improving the yields and net income from maize–bush bean intercropping combinations.

Maize and Rice Intercropped and Followed by Beans. Intercropping of maize and rice is an important practice in the lower elevations above the Atlantic coastal plain in Honduras. Because of a high degree of variability in rainfall amount and pattern in this area, the combination of these two basic food cereals provides insurance against crop failure and improves average net income. Comparison has indicated substantial increase in land productivity and potential net income of maize intercropped with several rows of rice and followed by monocrop cowpeas over monocrops of corn or rice followed by common bean (CATIE, 1979).

Maize Intercropped with Squash. Intercropping maize with squash is one of the very old practices of Central America, dating from pre-colonial times, and it continues as a means of increasing land productivity at intermediate to high elevations. At Yojoa, Honduras, the substitution of pipian (a squash with high market value) for ordinary squash may increase the net income per unit of land substantially (CATIE, 1979).

Maize Intercropped with Sorghum. In intermediate elevation areas with long dry seasons, the intercropping of maize with sorghum is an established practice, which has provided insurance against crop failure in unusually dry years. In Nicaragua and El Salvador CATIE farmer trials of this association demonstrated increased land productivity and net income compared with monocropping. Improvements in management practices increased both maize and sorghum yields and net income from the crop.

Maize Intercropped with Cassava and Beans. At low elevations in Central America intercropping maize and cassava increases land productivity and decreases risk of crop failure. Association of common bean with the cassava was studied by CATIE on farmer trials at Cariari, Costa Rica (1979).

Sorghum Intercropped with Beans. In the agricultural areas of Central America that have long dry seasons, sorghum may replace corn as the dominant grain cereal, or one or two crops of beans may be grown as monocrops. The latter is risky in the drier areas. CATIE (1979) has found in farmer trials at Samulali, Nicaragua, that intercropping sorghum and beans in a first planting, followed by a second planting of beans alone, increases both land productivity and net income and at the same time decreases soil erosion and water runoff. This modification of planting pattern, however, also changes the periods of peak demand for family labor and has to be weighed against the alternative uses for that labor.

Associations of Cropping Systems in a Farm Unit

Not only are there different associations of cropping systems within a bioclimatic zone, but the associations of cropping systems within a farm unit may differ markedly in response to different microclimates, the physical condition of the soil and its fertility, and the effects of terrain on potential soil and water loss.

Under climatic conditions of a well-defined dry season lasting three to five months, cropping systems with irrigation of arable land are frequently associated with permanent cultivation, with perennial crops under rain-fed conditions, or both. Usually irrigation is used because of limitations in the water supply, and the irrigated area occupies the land of lowest elevation. In Taiwan (Cheng, 1975), where the government policy has provided for the needs of the small farm rural population, the major share of the irrigation water goes to the western lowland rice area. The objective is to provide for two crops of paddy rice a year, which may be associated, in relay and succession, with a winter vegetable crop and a summer vegetable crop in between the rice crops. This very intense use of the land requires careful timing of planting and harvest during the four successive cropping periods. But in some irrigation districts there is only enough water for one rice crop, with enough supplementary irrigation to support a succeeding crop of sweet potatoes. At the higher terrains, a limited amount of irrigation water is available for a single rice crop, usually on terraces, which may be associated with separate areas under production of fruit or tea under rain-fed conditions, together with some winter vegetables. In these areas, it is not uncommon to find small herds of milk cows, whose milking stables are located above the level of vegetable gardening so the liquid manure may flow efficiently to the fields below.

In the Batanga area on the island of Luzon, Philippine Islands, under upland rolling conditions and a short dry season on a fertile soil, coconut palms, bananas, plantains, and coffee may be intercropped, and the intervening open lands are planted to upland rice, followed by vegetables for market.

In the intermediate elevations of Central America, with a dry season of three to five months, corn and common beans are either interplanted or planted in sequence as monocrops. In addition, some of the arable land may be left fallow to be grazed by cattle, and some coffee may be planted as a perennial crop.

From these few examples, it is possible to visualize the large array of possible associations of crops within cropping systems and cropping systems within farming systems.

Recapitulation

In this chapter, we have characterized and discussed the major cultivation systems of the tropics largely in terms of the intensity of land use, the ecological zoning of crops by their climatic requirements, the associations of cropping systems with ecological zones largely according to the social and economic infrastructure, and the identification of cropping systems by dominant crops. We have also provided illustrations of important cropping systems that have evolved in East Asia, West Africa, and Central America.

It should be clear from this review that systematic cataloging of cropping systems cannot be expected to cover all of the interactions among crops, ecology, and economic and social environment that determine the systems appropriate to specific sites.

But we can see that the cropping systems used by small farmers on specific sites of the tropics are usually appropriate to the local land-use intensity requirements, that the individual crops grown are reasonably well-adapted to the ecology of the site, that associations of cropping systems within sites and on individual farms develop in response to the local social and economic infrastructure, and that cropping systems develop from and around the dominate crops within them. The small farmer's choices are made within this context. The success or failure of technological changes or innovations that deviate from the existing systems depends on how well they complement those systems and enhance their opportunities to satisfy the objectives of the farmers.

Farming Systems, Including Animals

Since the animal components of farming systems are dependent for food on edible plants and their products, and since the mobility of animals allows them to be adapted to wider ecological ranges than crops, they will be considered within the context of integrated crop and animal farming systems.[1]

Developing countries account for 60 percent of the world's populations of domesticated ruminants, swine, and poultry. Because of ecological and socioeconomic constraints, however, no more than one-fifth of the world supply of edible animal products comes from these countries (FAO, 1979). Most of these animals are husbanded by small farmers and herdsmen whose production methods are not and cannot be guided by simple transfer of technology from the more developed countries. One important reason is that cereal grain production in the LDCs is either inadequate for or barely enough to satisfy minimum needs for human consumption. Nevertheless, there is a high degree of interdependence of humans and animals in these subsistence regions.

The priorities for livestock ownership in developing countries are different from those in industrialized nations. Reduction of risk, accumulation of capital, and the provision of power (traction), fertilizer, and fuel frequently have higher priority than the provision of food and generation of income. Social status and cultural needs may also be relatively high in the order of priorities. Despite the relative mobility of animals, there is wide variability in their role and importance and in the dominant species and breeds in the farming systems of different regions. Thus pastoral herding of cattle is dominant at high elevations of

[1]Our analysis depends heavily upon the working papers of a conference on the subject held in Bellagio, Italy, on October 19, 1978 (McDowell and Hildebrand, 1980), as well as on several papers by McDowell (1977, 1978, 1979, 1980). Tables and figures are drawn from McDowell and Hildebrand (1980).

eastern Africa, and sheep and alpaca are dominant in the Andean puna of South America; ducks and fish culture may be part of the small farming systems in some humid tropical regions of Asia, and the water buffalo and Zebu breeds of cattle are the dominant ruminants in vast areas of the wet tropics of Southeast and South Asia. Swine and poultry, on the other hand, together with a few cattle, are found in the small farm systems over a very wide range of ecological situations in many regions.

Prevailing Farming Systems by Regions

Several scholars have attempted to identify or systemize the prevailing farming systems of regions and of the world (Grigg, 1974; Kolars and Bell, 1975; Whittlesey, 1976). These classifications have been done on several bases, including geography (political and physical), climate, type of crop or animal, and the production method for that species. For the purposes of this discussion, we believe that farming systems can be more readily understood if we focus on crop-animal interactions.

Tables 7.1, 7.2, and 7.3 identify and characterize the prevailing systems employed on small farms in Asia, Africa, and Latin America, with the dominant crops, the predominant animal species on the farms, and the main feed resources for the animals. A farming system consists of a small number of major or dominant crops and numerous minor crops. The systems given attention here are those with an animal complement, so that dominant crops largely determine the feed source and are a major factor in selecting animals. Nutrient flow through the system is critical in limited-resource agriculture, and crop-animal relationships are critical to its efficiency. Crop-animal relationships as well as requirements for social organization have implications for labor use. For instance, village security and social structure largely determine the way in which animals are tended. Market structure must also be aligned to needs of the farming system.

These and many additional factors describe the complex of interrelated physical, environmental, and social elements that must interact in any particular system. In order to understand mixed farming systems in small farm agriculture, one should first look at a particular crop-animal interaction and be familiar with its essential elements. Then one can look at the range of conditions under which it is found. A final step to understand change in the system across environments. The classification proposed here is not so much intended to present new information on the specific systems as to suggest a conceptual framework to guide further study.

As an example, consider the coastal fishing and farming complexes in Asia (Table 7.1). These systems are found across most countries of Asia and also represent the predominant systems in the smaller islands across the Pacific. They are adapted to areas of relatively high population density and are found on the extremely poor soils of the coastal areas. These systems are designed for intensive use of the scarce resources in the coastal environment. The major crops, determined to a large extent by soil type and fertility, are coconuts, cassava, cacao and rice. The coconut by-products are used for swine feed; the marine by-products, such as fish trimmings, shrimp, or nonmarketable marine products taken along with the commercial catch, are fed to ducks. Cattle and goats are pastured under the coconut palms or in the more marginal land extending back onto the slopes of the hills, which are usually not far from the coast. The coastal fishing and farming complexes are highly specific to the physical and geographical environments in which they are found, but since these environments spread across the full length of Asia and Oceania, the system transects an extremely broad socioeconomic range.

If one understands the interaction of the system in the coastal area of southern Luzon in the Philippines, it will seem familiar wherever the same conditions are found. The selection of animals to match food availability, the matching of crops to their specific low-fertility environment, the use of animals to concentrate nutrients for cycling into the limited but all-important food-crop areas, the suitability of animals and food crops for marketing over long distances, the high diversity of enterprises within the system, giving it both biological and economic stability, are all crucial points in understanding its function. The system is, in cases of extreme isolation, ideally suited to subsistence conditions. Where resources are somewhat more plentiful and markets available, the system becomes immediately commercialized. It is relatively self-sufficient and self-sustaining, requiring few new inputs and a minimum or rural infrastructure.

Each of the nine other farming systems listed for Asia (Table 7.1), the ten for Africa (Table 7.2), and the four for Latin America (Table 7.3) could be described in a similar manner, studying social adaptability, biological stability, economic stability, nutrient recycling or energy-flow characteristics, infrastructure required, adaptability to commercialization, or a host of relevant features. This approach should be used not only to study and appreciate the complexity of farming systems but also to structure research and development strategies for those systems. The major advantage of the approach is that it increases the probability that the technology derived can become immediately adapted to the situations into which it is to fit. Such an approach minimizes the risk of developing a new and potentially productive technology that would be unacceptable because it did not fit into the farming system for which it was intended. If a lack of fit occurs, the reason for nonacceptance is usually a net reduction in productivity of the system because interactions among components are not adequately understood by the technology developers.

The two subsequent sections of this chapter are intended to aid in understanding the complexity of small farm systems and the delicate balances inherent in the interdependent nature of crops and animals.

Characteristics of Four Selected Systems

The objective of this section is to direct attention to various levels of integration of crops and animals and portray the infrastructural dependence within selected systems. We discuss four important prevailing systems, using a standard format for ease in comparisons.

In figures 7.1, 7.2, 7.3, and 7.4 the box identified as "Market" represents all off-farm activities and resources; hence it includes products sold or labor going off the farm as well as purchased inputs and household items. The "Household" is the core of the farm unit. In preparing the models of the systems, labor use, sources of human food, fuel, household income, animal feed, and the roles of animals were the main focus. The solid arrows ($\rightarrow$) depict strong flows or linkages (for example, more than 20 percent of total income arises form the sale of crops, animals, or household-processed products). Broken arrows ($---\rightarrow$) show that sales of crops or animals contributed less than 20 percent of household income, the interchange among functions was intermittent, or no routine pattern was identifiable; for example, the shifting cultivation (swidden) farmer of Southeast Asia (Figure 7.1) visits the market only occasionally with no predictable pattern. Family labor applied on the farm is identified, but off-farm employment or the amount of hired labor cannot be quantified except generally and is indicated by broken or solid arrows.

For most products there is a direct relation to market, absent when little is sold or when the household changes the characteristics of the product before sale (for example, wool to yarn, milk to cheese, or

Table 7.1. Prevailing systems of agriculture on small farms, major crops and animals, main regions, and feed sources for animals of Asia

Farming system	Major crops	Major animals	Main regions	Feed sources
Coastal fishing and farming complexes, livestock relatively important	Coconuts, cassava, cacao, rice	Swine	P, T	Coconut by-products, rice bran
		Ducks	TW, T, M, P, I	Marine products, rice bran
		Cattle and goats	SL, P, M, I	Pastured with coconuts
Low elevation, intensive vegetable and swine, livestock important	Vegetables	Swine	C, TW, HK	Sweet potato residues, rice bran, fermented residues from vegetable crops
		Ducks	HK	Crop residues, imported feeds
		Swine, fish	TW, M	Crop residues, rice bran
Highland vegetables and mixed cropping (intensive), livestock important	Vegetables, rice, sugarcane, sweet potatoes, Irish potatoes	Buffalo, cattle	P, T	Crop residues, rice bran, cut forage, sugarcane tops
	Vegetables	Sheep, goats	I	Crop residues, waste vegetables
		Swine	P	Crop residues
	Rice	Cattle, buffalo	Asia	
Upland crops of semiarid tropics, livestock important	Maize, cassava, sorghum, kenaf, wheat, millet, pulses, oilseeds, peanuts	Cattle, buffalo, goats, sheep, poultry, swine	IN, T	Bran, oilseed cake, straw, stovers, vines, hulls, hay

System	Crops	Livestock	Region	Feed resources
Humid uplands, livestock important	Rice, maize, cassava, wheat, kenaf, sorghum, beans	Swine, poultry, cattle, buffalo	Asia (>1,000 mm rain)	Stover, weeds, crop by-products, sugarcane tops
	Sugarcane	Cattle, buffalo	T, P, I	Sugarcane tops, crop residues
Lowland rice, intensive livestock	Rice, vegetables, pulses, chick peas, mung bean, sugarcane	Cattle, buffalo, swine, ducks, fish	Asia	Crop residues, weeds, crop byproducts, sugarcane tops
Multistory (perennial mixtures), livestock some importance	Coconuts, cassava, bananas, mangoes, coffee	Cattle, goats, sheep	P, IN	Cut and carry feeds from croplands
	Pineapple	Cattle	P, I	Crop residue, byproducts
Tree crops (mixed orchard and rubber), livestock some importance	Orchard, trees, rubber, oil palm	Cattle, goats, swine	P, M, South T	Grazing or cut and carry
Swidden, livestock important	Maize, rice, beans, peanuts, vegetables	Swine, poultry, goats, sheep	Asia	Animals scavenge
Animal-based	Fodder crops	Cattle, buffalo, goats, sheep	I, M, IN	Cut and carry fodder, crop residue

Key to main regions: C, China; HK, Hong Kong; IN, India; I, Indonesia; M, Malaysia; P, Philippines; SL, Sri Lanka; TW, Taiwan; T, Thailand.

Table 7.2. Prevailing systems of agriculture on small farms, major crops and animals main regions, and feed sources for animals of Africa

Farming system	Major crops	Major animals	Main regions	Feed sources
Pastoral herding (Phase I, L = >10) animals very important (symbiotic relationships)	*Vegetables* (compound)[a]	Cattle, goats, sheep	Savanna (southern Guinea)	Natural rangelands, tree forage
	Millet, vegetables	Cattle, goats, sheep	Savanna (northern guinea and Sahel)	Natural rangelands, tree forage, crop residues
Bush fallow (shifting cultivation, Phase II, L = 5–10), animals not important	*Rice/yams/plantains* maize, cassava, vegetables, tree crops, cocoyams	Goats, sheep	Humid tropics	Fallow, crop residues
	Sorghum/millet maize, sesame, soybeans, cassava, sugarcane, tree crops, cowpeas, vegetables, yams	Cattle, goats, sheep, poultry, horses	Transition forest/savanna Southern Guinea Northern Guinea and Sahel	Fallow, straws, stover, vines, cull roots, sesame cake
Rudimentary sedentary agriculture (shifting cultivation, Phase III, L = 2–4), animals important	*Rice/yams/plantains* maize, cassava, vegetables, tree crops, cocoyams	Goats, sheep, poultry, swine	Humid tropics	Rice bran, cull roots, straws, crop residues, vines, stover

Farming system	Crops	Livestock	Ecological zone	Feed resources
Compound farming and intensive subsistence agriculture (shifting cultivation, Phase IV, L = <2), animals important	*Sorghum/millet* maize, sesame, cotton, sugarcane, tree crops, cowpeas, yams, tobacco, groundnuts, vegetables	Cattle, goats, sheep, poultry	Transition forest/savanna Savanna (Guinea and Sahel)	Stover, vines, sugarcane tops, cull roots, or tubers, tree forage, groundnut cake, brans
	Rice/yams/plantains maize, cassava, vegetables, tree crops, cocoyams, yams	Goats, sheep, swine, poultry	Humid tropics	Rice straw, rice bran, vegetable waste, fallow, vines, cull tubers or roots, stover, tree-crop by-products, palm oil cake
	Vegetables sugarcane, tobacco, sesame, maize, tree crops, groundnuts	Goats, sheep, poultry, swine	Transition forest/savanna	Vines, stover, tree-crop by-products, groundnut cake
	Vegetables/millet cassava, cowpeas, tobacco, cotton, groundnuts, tree crops[b]	Cattle	Savanna (Guinea and Sahel)	Vines, tree-crop by-products, cassava leaves, fallow
Highland agriculture, animals important	*Rice/yams/plantains* maize, cassava, vegetables, plantain, cocoyamas	Goats, sheep, poultry, swine	Humid tropics	Fallow, leaves, stover, rice by-products, cull tubers, cassava leaves, vegetables residues
	Sorghum soybeans, cowpeas, cassava, maize, millet, groundnuts	Cattle, goats, sheep, poultry	Transition forest/savanna	Stover, vines, groundnut cake

(continued)

Table 7.2 (Continued)

Farming system	Major crops	Major animals	Main regions	Feed sources
Flood land and valley bottom agriculture, animals of some importance	*Millet/sorghum* maize, groundnuts, cowpeas, sesame, tobacco, cotton, vegetables, cassava, yams	Cattle, goats, sheep, poultry, horses, donkeys	Savanna (Guinea and Sahel)	Crop residues, some oil cake, brans, stover, vines, cull tubers
	Rice/yams/plantains maize, vegetables, sugarcane, rice, yams, cocoyams, millet, groundnuts	Goats, poultry	Humid tropics	Crop residues, vines, grazing
	Rice vegetables, maize, millet, groundnuts plantain, sugarcane, cocoyams	Cattle, goats, sheep, poultry, swine, horses, donkeys	Transition forest/savanna	Straw, stover, molasses, brans, groundnut cake
	Yams/sugarcane maize, cowpeas, cocoyams, groundnuts, vegetables, plantains, rice, yams	Cattle, goats, sheep, poultry, swine, horses, donkeys	Savanna (Guinea and Sahel)	Vines, brans, cull tubers, molasses, sugarcane tops
Mixed farming (farm size variable), animals important	*Rice/yams/plantains*	Two or more species (widely variable)	Humid tropics	Fallow, straw, brans, vines

Farming system	Crops	Animals	Ecological zone	Feed resources
	Rice/vegetables yams, cocoyams	Some cattle	Transition forest/savanna	Fallow, vines, straw
	Sorghum/millet groundnuts, cotton, tobacco, maize, cowpeas, vegetables	Cattle, goats, sheep, poultry, horses, donkeys, camels	Savanna (Guinea and Sahel)	Stover, vines, fallow
Plantation crops, East Africa (small holdings), animals of some importance	*Coconuts* vegetables, maize, plantains, cocoyams, cassava	Cattle, horses, donkeys	Humid tropics / Transition forest/savanna	Grazing or cut and carry
Plantation crops, (compound farms), animals of some importance	*Cacao* vegetables, maize, plantains	Goats, sheep, poultry, swine	Humid tropics	Grazing or cut and carry, stover
	Tree crops sugarcane, plantains	Goats, sheep, poultry, swine	Transition forest/savanna	Grazing or cut and carry, sugarcane tops
Market gardening animals may or may not be present	*Vegetables*[b]	Variable	Humid tropics / Transition forest/savanna	Natural rangelands, crop residues, browse plants, range forbs

Key to abbreviations under farming systems: L = C + F/C; L, land-use factor; C, area of cultivation; F, area in fallow.
[a]Enclosed areas around household or village.
[b]Present or absent, depends on area.

Table 7.3. Prevailing systems of agriculture on small farms, major crops and animals, main regions, and feed sources for animals of Latin America

Farming system	Major crops	Major animals	Main regions	Feed sources
Perennial mixtures large farms; livestock relatively unimportant	Coconuts, coffee, cacao, plantains, bananas, oil palm, sugarcane, rubber	Cattle, swine	All	Natural pastures, crop by-products, cull material
Commercial annual crops medium to large farms, livestock moderately important	Rice, maize, sorghum, soybeans, small grains	Swine, cattle, poultry	All except CI	Pasture, crop residues, grain
Commercial livestock a. Extensive Large to very large, livestock dominant	None are important	Cattle (beef)	C, V, Br, Bo, G, CA	Natural grasslands
b. Intensive Medium to large, livestock dominant	Improved pasture, some grains	Cattle (dairy), swine, poultry	All	Natural and improved pasture, feed grains, crop by-products
Mixed cropping Small size in settled areas Medium size in frontier areas Subsistence or nonetized economy Livestock relatively important	Rice, maize, sorghum, beans, wheat, cacao, plantains, coffee, tobacco	Cattle, poultry, goats, sheep, donkeys, horses, mules, swine	All	Natural pastures, crop residues, cut feed

Key to main regions: All, all countries; Bo, Bolivia; Br, Brazil; C, Colombia; CA, Central America; CI, Caribbean Islands; V, Venezuela.

manure to dung cakes). Household modification is shown by solid arrows from crop or animal products through household to market. Even though all crops require some processing, a distinction is indicated only when the household changed an already marketable product.

Fuel is extremely important on small farms. Gathering of wood or other materials often constitutes a significant expenditure of labor or may represent an important source of income. In each system, the major fuel sources were identified.

The four models shown in these figures are examples of major systems. Hundreds of models would be needed to characterize all small farm systems. Through an appreciation of the "interaction effects," however, the rationale of the "whole system" on small farms can be better understood and helps to explain why a single phase of technology, such as a new variety of maize, may be rejected by small farmers.

Shifting Cultivation in Asia. The shifting cultivation (swidden) system (Figure 7.1) is employed on 30 to 40 percent of all land in tropical Asia (Grigg, 1974; Harwood, 1979). It is found in widely dispersed settlements employing shifting agriculture and slash-and-burn technology. A family or household cultivates approximately two hectares per year using manual labor. The main implements are hoe and dibble (planting) stick. Plant residues are usually left in the fields for mulch. Each family has pigs and chickens without controlled management (scavengers), thus there is no systematic recycling of nutrients, although some manure may be retrieved for certain crops near the household. There is a complex interplanting of annual crops and few perennial crops. Two to four years of cropping are followed by an extended fallow period. There may be little animal-crop competition because the fields are ordinarily several hundred meters or more from the village, where the families reside. Fuel is a minor problem in this system because of low population densities and the presence of forest or fallow.

Farm infrastructure is low in that few capital inputs and services come from outside the village. Mutual assistance within the village is the main source of aid. There is no systematic plan for sale of livestock or identifiable pattern of service use for animals. Most sales of animals are for emergency needs, with the greater proportion being consumed to celebrate cultural and religious events (DeBoer and Weisblat, 1980).

The soils are generally marginal in fertility and on moderate to steep slopes, so erosion is often a serious problem. Wildlife from forest fallow areas often prey on crops or even on the small animals.

The system has several assets. The usually low population pressures

Figure 7.1. Shifting cultivation (swidden) farming system in Asia: Shifting agriculture, low integration of crops and animals (animals free-roving or tethered)

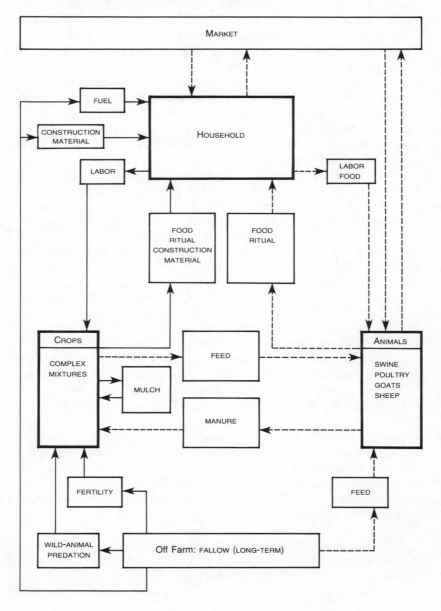

permit long-term fallow. Diversified cropping is widely practiced and enhances soil conservation. The constant shortage of labor slows expansion of cultivation and thereby decreases risk of erosion. On the other hand, the system has serious liabilities, such as poor access to markets and inadequate power for tillage or transport. Increasing land pressure caused by population growth or expansion of permanent ranching and timber harvest causes the fallow system to break down in many areas (Harwood, 1979).

The opportunities for improving shifting cultivation (swidden) systems are often good. Returns from crops and environmental stability could be improved through the use of perennial crops, bunded paddies (earth barriers to retain water on the land), terraces, and planned grazing areas so that buffalo or cattle can be incorporated into the system. Use of large ruminants would improve the opportunity to accumulate capital. These changes may require development of appropriate technology and guidance and may depend on a shift in attitude on the part of policy makers, most of whom see the swidden system as wasteful and making little contribution to agricultural production.

Humid-Upland System in Asia. The upland system (Figure 7.2) is widespread over the humid tropics of Asia. There are well-developed farmsteads with permanent, cleared fields but with no bunding and no irrigation. The major crops are rice, maize, cassava, wheat, kenaf, sorghum, and beans. Most households have small numbers of several species of animals. Swine and poultry are prevalent, followed in popularity by cattle and buffalo. Sheep and goat numbers are normally low. Where tall-growing crops (maize and sorghum) are cultivated, cattle are kept to use crop residues. In rice areas buffalo predominate. Frequently, one or two buffalo or cattle are kept for use in land preparation and to provide transport for crops, crop residues, and to some extent members of the family. Swine are tethered or penned, and cattle or buffalo are tethered at night so that manures may be collected and to avoid theft. The manures are frequently composted with crop residues. Poultry are usually free-roving.

Fuel is not yet a severe problem in much of the area devoted to the humid-upland systems but is becoming scarcer as more of the forests are cleared.

The farm infrastructure is variable, developed for some areas but extremely limited for others. Land tenure and social services are also variable. Many upland areas are distant from markets.

The land ranges from rolling hills to steep slopes. The soils are moderately fertile, and in general drainage is good. Erosion hazards are usually moderate. The rainfall is seasonal and may be erratic during the rainy season, thus in some areas periods of moisture stress are frequent.

Figure 7.2. Humid-upland farming systems in Asia: Permanent cropping, moderate integration of crops and animals (animals tethered or herded)

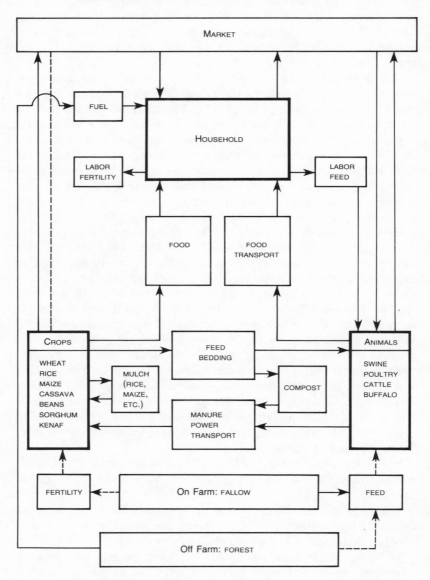

Among the assets of this system are the opportunity for multiple cropping, excellent potential for crop and animal integration, good potential for small-holder dairying with crop rotation, and feasibility of cooperative production and marketing. Rice is milled at the village level; therefore, rice bran and other by-products are available for supplementary feeding of animals. Some of the current limitations to increased output are inadequacy of credit and animal health services, insufficient power for tillage, and limited access to markets. In addition, farms are often so geographically fragmented that the potential for grazing is limited. Considering the assets and liabilities, the potential appears good for improvement through increased cropping intensity, especially of fodder crops for animal feeding; increased animal holdings so that farmers could have increased scheduled outputs for marketing; expanded farm infrastructure; use of draft power; and greater milk production.

With time, the upland areas of Asia promise to meet the rising demand for milk and meat through greater crop-animal integration (DeBoer and Weisblat, 1980). Crop-animal integration on small farms minimizes the need for feed concentrates in animal production, and there is some possibility for on-farm self-sufficiency in power (gasohol, biogas) based on conversion of sweet potatoes and cassava.

Lowland Rice System in Asia. The lowland rice system (Figure 7.3) is characteristic of traditional small-farm operations in the river valleys, including first and second terraces above the valley bottoms, and in coastal areas of Asia, including southern China. These areas have at least three months of rainfall above 200 mm and a dry season of two to six months. Length of dry season is a major factor in feeding animals. The areas are tropical (frost-free). Both human and animal population is dense. Rice is the major crop, followed in importance by garden vegetables and food legume crops. The use of fertilizer and manure assures high crop yields. Rice is milled in the villages; therefore, rice bran and other by-products are available. Rice bran has a good level of crude protein (12 to 15 percent) and a significant amount of oil or fat; hence rice culture–livestock integration adds to the intensification of this farming system (Maner, 1980).

Animals provide income and manure as well as fuel in southern Asia. The major species are cattle, buffalo (swamp or carabao), swine, chickens, ducks, and geese. The bovines are kept to eat crop residues and to supply manure and power for tillage and transport. Old draft animals are sold for meat. Rice by-products and cut grass are fed to swine. The pigs are sold for additional income. The ducks and geese feed on grains lost during harvest and on insects and weeds in and around the irrigation canals. Most of the eggs and meat from chickens,

Figure 7.3. Lowland rice farming system in Asia: Permanent cropping, high integration of crops and animals (animals confined)

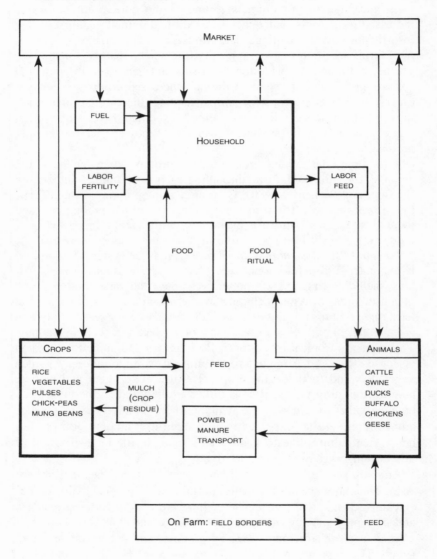

ducks, and geese are consumed within the household or in the immediate community. The farms are small and fragmented, which makes control of grazing animals difficult. As a result, the larger livestock are confined and hand-fed, which permits collection of manures. Another reason for tethering or confinement is security; theft of animals is a problem. Animals, especially the buffalo, are a strong feature of the cultural system for ritual use (Barnett, 1980).

Because of high population pressures, no land is available for producing fuel. The high rate of use of manures on crops precludes them as a source of fuel. Hence, in this system, the primary source of fuel is kerosene purchased at the market.

The assets of the lowland rice systems are numerous. Multiple cropping can be expanded to reduce dependence on a single crop (Riley, 1980). Farmers are experienced in the care of animals. Labor for livestock production is plentiful during long periods. Irrigation serves to reduce risks in cropping, thus farm capital is relatively easy to accumulate.

There are certain restrictions to expansion of crop and livestock production. For example, the nutritive value of straw of the new, high-yielding varieties of rice is lower than that of the traditional varieties, and draft animals may require supplementary feed to maintain their work efficiency (Goe and McDowell, 1981). Multiple cropping reduces the amount of grasses and weeds traditionally cut and fed to animals. Irrigation and multicropping may increase the value of labor to such an extent that interest in livestock declines (Riley, 1980; Plucknett, 1979). Increased use of pesticides and herbicides in multicropping may limit fish and duck production in rice paddies. Increased mechanized harvesting may cause rice milling to shift away from the villages, and thus stimulate development of large commercial livestock operations that could monopolize markets.

On the whole, the intensity and efficiency of crop-livestock (non-ruminants) production are higher on small farms in the lowland rice system than in the other systems described in this chapter (Maner, 1980). Even so, there is good potential for improvement. For example, fertilizer costs could be reduced by cropping of legumes following rice harvest, to take advantage of residual moisture in the paddies. The legumes would complement low-quality rice straws for livestock feeding. Other approaches that involve institutional changes include securing land tenure to encourage accumulation of animals; introducing long-term use of forage legumes for animal production; adopting a multidisciplinary research and development approach to maximize farm income; supplying market assistance to small-scale swine, chicken, and duck producers in order to overcome the high unit cost of marketing small numbers of animals; and offering credit and extension services on a year-round basis.

Tree-Crop Farming. Perennial tree crops (Figure 7.4) such as coconuts occupy land for as long as fifty years. Trees are spaced 8 or 10 m or more apart, leaving large light-exposed surfaces that can be used for cropping or grazing, especially when the trees are immature.

Coconut and oil-palm producers, among others, have the problem of managing the shaded ground areas. Competition from annual and perennial weeds is continuous. Farmers initiate animal production by tethering or fencing animals under the trees. Grazing improves weed control, enhances nut collection, and results in some benefits in nut production from the manure deposited. Without animals for grazing, farmers must control weeds by hand, cutlassing four to five times per year, or intercropping with food and cash crops.

Of the more than 6 million hectares of coconuts in the world, over 90 percent are found in tropical Asia. The grazing of cover crops under coconuts is practiced widely not only in Asia but also in East Africa and the Caribbean Islands (Plucknett, 1979). More than 90 percent of the world's coconuts are grown by small holders to whom the coconut-animal system is of special importance. The system can be employed with other trees crops, such as cashew and rubber, but it is not practical with low-growing trees or bushes such as coffee, tea, or cacao.

Where tree cropping with livestock is practiced, the level of integration is low to moderate because few of the by-products of the tree crops are available for animal feed. Copra meal, for example, can be a good feed, but the oil processing does not occur in villages. The livestock are usually tethered among the trees or, in the case of swine, permitted to rove or scavenge. Cattle may be used to transport the crop from the farm during certain seasons, but there is no consistent pattern. Manure is not collected, and any milk produced generally goes for home consumption. The animals are, therefore, principally a means to generate capital and reduce risk. Fuel for household use is generally a problem. Tree crops and prunings are for the most part poor for burning; hence fuel must be bought or sought in forests some distance from the farms.

The potential for expansion of the integration of animals and tree-crop farming is excellent. There is renewed government interest in expanding crop production, especially of coconuts. Research results on better tree density, on the benefits of fertilizer application to increase production, and on the complementarity of forage legumes are becoming available. Labor for animal care is plentiful during long periods.

There are also certain limitations to expansion of animals on the tree farms. For example, the market for livestock products must be good

Figure 7.4. Tree-crop farming in Asia: Long-term cropping, low integration of crops and animals (animals tethered or roving)

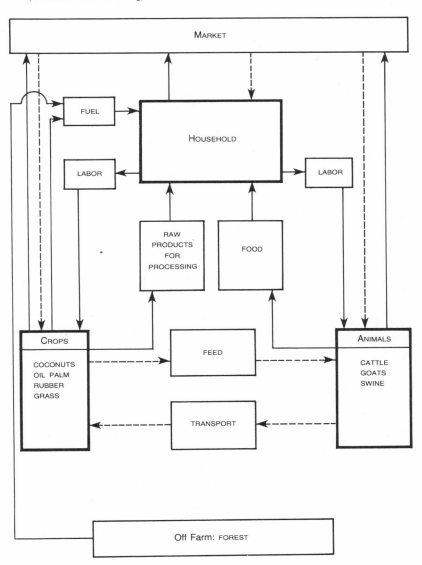

enough to persuade farmers to make new inputs, such as planting improved grasses or legumes. A higher return will be required to offset possible reduced yields from the trees because tree density must be lower to assure growth of forages (Plucknett, 1979).

Conclusions

The examples cited above show that small farm systems are highly variable; they are complex and require good managerial skills to operate effectively. Appreciation of the interdependence of cropping and production from animals permits technicians to understand the small farmer's rejection of some technological innovations because of the risk of creating an unacceptable imbalance in the whole system. Use of an improved variety of maize, for example, may decrease stover (crop residue) yield to a point of inadequacy for animal feeding. The illustrations given make it clear that there is a need for technology more suitable to small farm systems (Maner, 1980).

Two important functions not portrayed in the diagrams are that animals are valuable during times of food or cash shortage and that animals can act as a buffer against contingencies such as illness, accident, famine, seasonal food shortage, or the need to help relatives in time of trouble.

Technological solutions for either cropping or animal production are not value neutral. Within a household or in a community some will gain from the use of technology and others will lose. The complexity of the systems described, as well as the need for determining net benefit, show the need for representatives of various disciplines to interact to assess the potential impact of technology. The need for introduction of a third dimension—social factors—into mixed farm systems has become a feature of the research program of the International Livestock Centre for Africa (ILCA). The ILCA field teams in the highland, semiarid, subhumid, and humid zones of Africa all include both specialists in crop and livestock production and sociologists, anthropologists, and economists. Preliminary results show that interaction among disciplines aids in understanding interrelationships within the production system and in screening interventions (Barnett, 1980).

8

Tools and Machines

Agricultural machines are specific and easily identified. They have been hailed as essential to increasing food production in food-short, underdeveloped countries; they have also been cursed as a waste of capital and not in the best interest of the poorer strata of society. This chapter will not attempt to determine whether machines are good or evil because such a broad question is unanswerable. Rather, the question addressed is whether a specific implement improves the specific farming system in which it is used.

In North America, where agricultural mechanization today is primarily used to replace labor, we associate machinery with labor displacement and thus may reasonably ask if mechanization is desirable in a poor and primarily agricultural country that experiences underemployment during most of the year. Crops are produced by the interaction of sunlight, water, and nutrients. By cultivating the soil, people use energy to increase yield by creating better conditions for crop growth; but a small plot cultivated by machine may produce no more than the same plot cultivated by hand. Why, then, is mechanization considered as a means for increasing food production?

We should begin by considering the plateaus of energy available to a farmer and the meaning of work and power. The first point to be considered is tillage. In an engineering sense, about the same amount of work is required to break and to pulverize the soil to create a seedbed whether done by a human fueled by rice, an ox fueled by forage, or a tractor fueled by petroleum. The rate at which the work is accomplished is determined by power available. Power is the rate of doing work. A man is rated at 0.1 horsepower (hp); an ox at about .7 hp; and a well-fed draft horse at 1 hp.

Horsepower per hectare is often used as an index of the viability of a country's agriculture. G. W. Giles (1967) suggested that field work requires a minimum of 0.5 to 0.8 hp per cultivated hectare in order to obtain satisfactory crop production. The sources of horsepower include man, animals, electric motors, and internal combustion engines. In another study Giles (1975) analyzed agricultural viability by using the average aggregate yield of major crops (cereals, pulses, oilseeds, sugar, potatoes, onion, and tomatoes) per horsepower expended. This comparison of aggregate values provides an overall view of agricultural mechanization, but, like gross national product (GNP), the values do not provide very useful indexes to the situations in many developing countries. Horsepower per hectare and tons per horsepower do not provide much help to one responsible for putting together a farming system for a specific site.

The production of food through agricultural activities makes man responsible for a system over whose essential elements—wind, rain, temperature, and sunlight—he has little control. This production model differs from that in industry. The model of a factory such as an automobile plant or a paper manufactory provides the plant manager with a large measure of control over the system. Tools and machines do not give the farmer control of his system but do allow him to take advantage of certain aspects of the system and to guard against what he sees as faults or calamities produced by the system.

Characteristics of Mechanization Often Cited as Advantages

Mechanization is often cited as one of the essential elements required for increasing crop production in many countries. The advantages of mechanization under specific agronomic and social conditions are timeliness, uniqueness, inertness, and reduction of drudgery, but not all four desirable characteristics are necessarily associated with the introduction of a specific machine.

Timeliness. There is usually an optimal time to prepare the soil and to plant. Timing is especially critical in areas of very low and highly irregular rainfall where the crop will be adversely affected if not planted within a particular time period. Power—animal or mechanical—and appropriate implements enable the farmer to cultivate greater areas than if only human power and hand tools were available. If power were available by rental or cooperative ownership of animal or mechanical power in exchange for human labor, it would have to be there at the time needed for the critical land preparation or harvesting operations.

Timeliness is also important in tropical areas, where more than one crop can be grown per year if turnaround time between the harvesting of the first crop and the planting of the second can be reduced. Mechanical power shortening this time may allow two crops whereas manpower alone will not. To plow an acre of land for rice planting demands fifty hours when done by a man with a hoe, eight hours by water buffalo and a simple plow, five hours with a 5-hp pedestrian tractor, and one hour with a 45-hp tractor equipped with a rototiller. Thus shortening turnaround time can contribute to crop production. For example, a 125-day variety of irrigated rice, yielding approximately 3,000 kilos per hectare, represents an average of 24 kilos of rough rice per hectare for each day from seeding to harvest. Each day spent preparing land for the next crop or lost because the land is still occupied by the crop ready for harvest can be viewed as a loss of 24 kilos of rice per hectare.

Although the effect of timeliness can be readily demonstrated under research conditions, there are few good studies of farmers' actions in regard to timeliness of operations after implements and tractors are introduced into an area. Hans P. Binswanger has reviewed and analyzed studies of the effects of the introduction of farm tractors in Southeast Asia (1978). He concludes that "the tractor surveys fail to provide evidence that tractors are responsible for substantial increases in intensity, yields, timeliness, and gross returns on farms in India, Pakistan and Nepal. At best, such benefits may exist but are so small that they cannot be detected and statistically supported" (p. 73). In most of the studies cited by Binswanger, the lack of statistical evidence of timeliness may have been a result of absence of information or the economics of capacity utilization. For example, on both bullock-powered farms and tractor-powered farms, there may not have been extra power available at certain times of the year because the farmer could not afford to invest in a power source he would use for only a few days a year.

A. S. Kahlon in his study of the Punjab (tropical wet-dry climate) quantified the timeliness achieved by farmers in a field situation (cited in Binswanger, 1978:37). Agronomic conditions were such that there was a six-week period during which the farmer would experience little yield whether he planted in the sixth week or the first week of the planting season. Binswanger attributes the slight superiority in timeliness of farms with tractor power over farms using bullocks to the probability that the tractor farmers are better managers. This conclusion raises an interesting but as yet unanswered question, "Why were the best managers using tractors?"

Narpat S. Jodha's study of an unfavorable agronomic region revealed that in Rajasthan (tropical dry climate), tractor usage increased rapidly because the combination of sandy soils, scant rainfall, and no irrigation resulted in a safe sowing period of only five or six days (cited in Binswanger, 1978:30). Tractors do not require rest and may be operated at night, enabling farmers to plant crops within the agronomic timeliness "window." Furthermore, night operation enables the farmer to work the tractor a greater number of hours per year and to reduce the fixed tractor costs per hectare.

Timeliness of field operations is thus more important in some regions and for some crops than others.

Uniqueness. The use of animal or machine power and appropriate tools makes it possible to cultivate land that is too difficult for normal human efforts. For example, a study of farmers in Terenos, Mato Grosso, Brazil (tropical humid climate), by John M. Sanders and Vernon Ruttan, revealed that farmers were replacing animal power with mechanical power because the best time for tillage and planting normally occurs toward the end of the five- to seven-month dry season and the heavy soils in the region cannot be satisfactorily plowed with animal power before the rains begin (cited in Binswanger, et al., 1978:276–278). Toward the end of the dry season the work animals are in poor condition and the soil is hard. Tractor-driven plows can work the soil under these conditions, so the crop can be planted just before the onset of the rains. When using poorly fed animals for power in these heavy soils, it is necessary to depend upon the first rains to soften the soil for plowing. In this respect, it is important to remind ourselves that draft animals, like humans and unlike machines, cannot be worked long hours without rest. For example, in the conditions described above, oxen are generally worked only half-days. The Mato Grosso study revealed that the average farmer using animal power worked 6.5 hectares while the farmer using a tractor worked 11.2 hectares per year.

Inertness. Tractors consume no fuel when not being worked, whereas animals must be fed throughout the year.

In northern Thailand (tropical moist or wet-dry climate) John V. Dennis, Jr. studied the use of water buffalo and tractors in village agriculture. In comparing water buffalo with tractors in single cropping and multiple cropping and the relation of these elements to the social scene, he concluded that 5 percent of the cultivated land area is barely sufficient to provide forage for the water buffalo (1978). Under a system of one crop per year, fallow fields are available for grazing for nine months of the year. Therefore, the sustenance of the animals depends on the cropping system.

Disease can be a problem. In 1975 an outbreak of hoof-and-mouth disease in the tropical moist or wet-dry climate of the Philippines accompanied by government-backed loans for agricultural machinery produced a very rapid increase in the number of pedestrian tractors.

In equatorial Africa south of the Sahara, in a band from about 15°

north to 20° south latitude, the tsetse fly and the accompanying try-panosomiasis ("sleeping sickness") debilitates and kills thousands of humans and animals each year. Some cattle, such as the N'dama, have some tolerance to trypanosomias, but N'dama are small and not generally used for draft (Desowitz, 1977).

Reduction of Drudgery. As bus routes and transistor radios have increased the lure of urban life, many young people seek to escape what they regard as the drudgery of work on the farm through migrating to cities. With high rates of unemployment and underemployment in many cities, reality often fails to measure up to expectations, yet this discrepancy does not stem the flow of rural-urban migration. If the lure of the city is to be dimished, life on the farm must be made more attractive, and selective machanization to relieve some of the drudgery of farm work may be one important means to that end.

Some Cases of Mechanization

Examples abound of the inappropriate use of agricultural machinery in developing countries. Shortly after independence in 1957, Ghana embarked upon an extensive program of agricultural mechanization without benefit of feasibility studies (Gunkel, 1968). The Soviet Union supplied Ghana with approximately two thousand tractors, including tracklayers with exhaust-heated cabs designed for cold climates. The tractors were assigned to state farms and to private farms represented by the United Ghana Farmers Co-op Council. The tractors were unsuitable for Ghanaian conditions, and insufficient complementary equipment such as plows, harrows, and other accessories was supplied. The primary causes of the failure of this mechanization scheme, however, were that supporting repair and training services were inadequate. There was no incentive for drivers to use machines properly, and management of tractor scheduling and field layout was inadequate. Remains of this equipment fiasco still dot the Ghanaian landscape today.

Examples also abound of machines contributing to the betterment of the lives of those who work in agriculture. In Malaysia (tropical wet climate) small farmers successfully use tractor hire services (Chancellor, 1971). Four-wheel farm tractors and pedestrian tractors, usually owned by farmers, are hired out for custom tillage on the land of farmers who do not own tractors. In general, the tractor service has a value to the nonowning farmer of about one and a half times the rate charged. Extending the use of tractors by providing more than one driver per tractor is common. The four-wheel tractors operate about

1,040 hours per year, and the pedestrian tractors operate about 400 hours per year.

Using hypothetical standard values, Chancellor found through an economic analysis of tractor ownership in Malaysia that tractor owners were losing money yet tractor sales to new owners continued to increase. The study also revealed that almost every farmer hiring the tractor service used the time saved to produce additional income. Three-quarters of the farmers making use of tractors produced the additional income through agricultural intensification, which did not displace rural labor to urban areas.

Binswanger (1978) also notes that the Indian, Pakistan, and Nepal economic studies reveal that returns to tractor owners from agricultural operations are close to zero or even negative. Why do these farmers continue to purchase tractors? In some instances, government subsidies and rising wage rates provided economic stimuli. Tractors also enable the owner to gain tighter and more secure control over farm work than is possible with hired labor.

As a rule, benefit-cost studies have not considered the non-agricultural benefits of tractors and agricultural mechanization. In developing countries, farm tractors are often used for transport. From the 5-hp pedestrian tractor pulling a two-wheeled trailer in the Philippines to the four-wheeled tractor pulling a wagon load of bagged cement in Ghana, farm tractors are used for off-farm transport. Although the value of this work cannot be quantified, it is nevertheless a benefit to the owner.

In the Philippines the IRRI-designed axial-flow thresher (IRRI, 1977), has been produced and sold in great numbers (nearly 14,000 units from 1974 through 1979). Since the thresher powered by a 7.5-hp engine can thresh rice faster than a farmer working manually using a threshing rack, the engine-driven thresher can displace labor and has undoubtedly done so in many cases. In the Philippines, as in many countries, a threshing crew is employed by a farmer to cut the grain, bring it to the threshing area, and thresh it manually. Some of these crews are now renting the small engine-driven thresher to thresh the grain, thereby making the work less arduous and threshing the farmer's grain more speedily. The members of the crew formerly doing the threshing are now transporting sheaves to the threshers.

Agricultural Tools and Equipment Research

How should agricultural tools and machines for low-income countries be developed? How should research institutions address this problem?

In the United States, government supports research and development for crops and agronomic practices while private industry develops and produces agricultural machinery. This division of responsibilities has worked reasonably well because U.S. agriculture has provided a profitable market for the manufacturers of farm machinery. But even in the United States we cannot always assume that private industry by itself will produce the machines farmers need. For example, New York State is the second largest producer of grapes, yet a few years ago the growth of this sector was blocked by a labor bottleneck in hand harvesting. A research and development project at the state's land grant college produced a pilot model of a mechanical grape harvester. At this point, private industry invested in redesigning the model to fit requirements for its manufacturing, sales, and service organizations. Today 85 percent of the grapes produced in New York are machine harvested, and grape acreage is increasing.

If, even in the United States, private industry alone has not been able to meet all needs for farm machines, it is unrealistic to expect the private sector to do the job unaided in developing countries. The national market may not be large enough to support a strong farm machinery industry, and the imported machines are not likely to be well adapted to the needs of farmers in the developing country.

The production and distribution of goods requires the interaction of people and necessitates the creation of sales, distribution, and service networks. In North America the agricultural machinery manufacturers have extensive networks consisting of engineering, manufacturing, sales, and service organizations. In addition, private dealers sell to the farmers and provide repair parts and service. These networks provide an excellent information and experience feedback loop between farmers and manufacturers. The result is a constant improvement of machines in response to the needs of the farmers.

In most developing countries the only existing farm machinery networks are put together by importers. The importer sells goods to local traders or dealers who sell to farmers. But there is no feedback loop to deliver the farmers' complaints to the importer and back to the foreign manufacturer. Even if the feedback existed, the information sent back to the manufacturer probably would not result in changes to make the machine fit local conditions better. A poor country's purchases of a particular machine usually represent a very small percentage of the foreign firm's output and could not justify the designing of a specific model for which there is limited worldwide demand.

Because agricultural conditions—soil, weather, crops, and agronomic practices—are site-specific, tools and machines also need to be site-specific. Therefore, useful mechanization requires manufacturing or adapting imported machines to specific regional conditions.

How should research be conducted on agricultural tools and machines for farming systems found in LDCs? It must be understood that introducing a new element into a system creates a new system, which is not identical to the old. Tools and machines should be adapted and designed for specific climatic, soil, and social conditions in the same way that crop varieties are designed (bred) for specific environmental characteristics.

It is sometimes suggested that in a developing country the first order of engineering research applied to agriculture should be the improvement or invention of hand tools. Useful results might not follow unless at the same time other major components of the agricultural system were altered. The hand tools in common use today for the most part represent the cumulative efforts of thousands of unknown farmers in developing a particular tool to fit the soil, crop, energy, capital, and social constraints of a particular cropping system.

Local improvements do continue. For example, blacksmiths in Ghana use broken auto or truck leaf springs to make better hoes, and in Java a blacksmith may be using old railroad rails to make better machetes. They may be unable to state that it is the higher carbon steel used for springs and rails which provides the harder, longer-lasting edges of tools made from these scrap materials. They learned by experimentation and trial and error. In most cases when foreigners have attempted to improve local tools, results have not been encouraging. The tools were often improved in the technical sense but perhaps at a higher cost, requiring imported materials or specialized training.

Although the development of hand tools may not justify high priority, the hand tools used by farmers should be studied in relation to the development of plant varieties and farming systems. For example, if a new variety of bean is to be used in a system where it would be threshed with hand flails, then during the selection and breeding program, the beans should be threshed by hand flailing instead of by a mechanized thresher to see if the variety is suitable. In developing a particular plant variety, it may be necessary to introduce a tool or a machine into a locality as part of the "package" in order to obtain advantages provided by the new variety for the farming system. When a tool or machine is introduced along with seed, fertilizer, and pest control, the tool or machine will then be accepted or rejected as a necessary part of a "package" in accordance with the cost/benefit to the farmer. In arriving at that judgment, the farmer will think not only of money but also of labor effort and status.

It is surprising to observe that tools and machines used extensively in one part of the world may be unknown to farmers elsewhere. It should be the responsibility of agricultural researchers to be alert to tools that have evolved in different areas of the world because such tools might be advantageous for cropping systems being developed for specific conditions under study. Agricultural extension effort to provide agronomic techniques and knowledge should also include training in the care and maintenance of tools and implements.

Tools and machines are usually distributed by importers and traders, and plant varieties are developed by government research stations, with little cooperation between the two. Furthermore, in many LDCs there is little intercourse between the engineering colleges and the agricultural colleges of the same institutions. In fact, engineering research may be under one ministry while agricultural research is under another.

Agriculture provides little prestige for engineers in many LDCs. Lack of communication with agricultural technologists limits the application of engineering principles to agriculture's problems. For example, at the Department of Mechanical Engineering of the Institute of Technology in Bandung, Indonesia, a class of mechanical engineering students designed and constructed an IRRI-type bellows pump. The students had been studying metals and metal fabrication, so they constructed the pump frame of steel, instead of using wood as in the IRRI model. Unfortunately, the pump was so heavy that a person could not easily carry it from paddy to paddy. Because of the need for low cost and manual transport, the pump would be unacceptable to Indonesian farmers. The students' design is not surprising because most of them had no practical agricultural experience.

There is a tendency in many developing countries to establish testing centers where imported or locally produced agricultural implements can be tested and compared and a judgment made as to whether license should be granted for the importation of specific machines. On paper, the scheme appears good, but it may fall short of expectations. First, the resulting bureaucracy may be slow-moving and dogmatic. Second, endurance testing done on a test stand often cannot simulate actual farm conditions and provide meaningful results. Third, the system provides the opportunity for bribery by importers anxious to receive certification. Fourth, a testing station by itself will not encourage the local production or modification of machines that are specifically designed for the system of agriculture used in the country.

Many scientists and financial donors do not view machine development as proper for a public research institute. This attitude is probably a reflection of North American practice, where plant breeding and agronomic research are funded by government and conducted in public research institutions but agricultural machinery research is funded and conducted by private companies. Nevertheless, several of the international agricultural research centers are making progress in developing mechanization and integrating it into the various farming systems being developed as noted below.

The International Crop Research Institute for Semi-Arid Tropics (ICRISAT) has concentrated its efforts on developing bullock-powered tools which enhance crop production in semiarid regions typical of ICRISAT's location near Hyderabad, India. It is important to note that animal-powered equipment is a component of the farming systems program.

The Centro Internacional de Agricultura Tropical (CIAT) in Colombia has performed some important research on the mobility of four-wheel farm tractors by use of oversized tires operated at low pressure as well as developing equipment for forming bunds (embankments of earth to keep water in a paddy). Some research has also been done in developing cassava lifters, but there is no formal agricultural engineering program.

The International Institute of Tropical Agriculture (IITA) in Ibadan, Nigeria, has a small agricultural engineering staff in its farming system program. Some thoughtful farming systems concepts have been developed, using mulch farming, herbicides applied with battery-powered sprayers, and no-till planting of maize and beans with manually operated planters.

The International Rice Research Institute has had an active group of engineers developing machines suitable for rice production on small farms. The pedestrian tractor (tiller), axial-flow thresher, and grain dryer are notable examples, as shown in Table 8.1.

Table 8.1. Commercial production of selected IRRI-designed machines, 1974–1979

	1974	1975	1976	1977	1978	1979
Pedestrian tractor	1,890	2,307	3,756	1,147	988	1,625
Axial-flow thresher	110	362	765	1,883	4,228	6,583
Grain dryer		233	1,100	1,675	119	137
Number of manufacturers		26	48	60	58	65

Source: Personal communication, April 5, 1980, from John A. McMennamy, IRRI, Los Baños, Philippines.

Three seldom-noted facts contributed to the successful adaptation of IRRI machines by farmers and the manufacturing of those machines by small manufacturers. First, the Philippines have a relatively well-educated populace and a great number of well-trained artisans and entrepreneurs in sheet metal manufacturing, welding, and machine repair. Second, engineers at IRRI spent much time in working with local manufacturers in building prototypes and in helping small manufacturers adapt IRRI designs to their specific shop equipment. These hours of consultation are not readily discernible in the IRRI annual reports. Third, Philippine government credit programs, high cost of imported equipment, and extension campaigns such as the Masagana 99 Program to increase rice production provided a climate for farmers to invest in locally made machinery for rice production.

The Ten-Horsepower Agriculture project in Egypt, sponsored by the Catholic Relief Services and the Ford Foundation, is an example of the successful planning and design of specific farm machinery to overcome labor shortages and increase land productivity through intensification of cropping (Koval and Behgat, 1980). An economic study of agriculture in the Nile Valley subsequent to land reform and the Aswan Dam revealed seasonal labor shortages in this irrigated region resulting from alternative employment of farm laborers at the times of planting, harvest, and threshing. Although the use of tractors has increased, their availability to small farmers is limited, and the high cost of fuel has resulted in increased use of animal power as well. A small diesel-engine-powered multicrop thresher-winnower and a low-lift water pump for irrigation were developed for local manufacture, and initial farmer acceptance has been satisfactory. Both relieve the demand for tractor power and can be powered by a transferable, standard, imported 10-hp diesel engine. The investment incentives for importers, manufacturers, and farmers (when given adequate credit from national banks) appear adequate to spur the adoption of these complementary tools. Research on design improvement is continuing. Examples such as these indicate that new machines, if especially designed to fit into indigenous farming systems, can make significant contributions to the welfare of small farmers.

9

Fitting the Parts Together

Though we are concerned in this book with meeting the interests and needs of small farmers, in Part II we have kept the farmers in the background while focusing on the physical and biological bases of the small-scale farming enterprise. Before moving farmers into the foreground in Part III, let us fit together the parts of the physical and biological environment which provide farmers with both opportunities and limitations.

Climate and Ecology

Tropical regions in less developed countries have been chosen as the source of illustrative materials because the majority of small farmers and most of the remaining land adapted to agriculture but not yet fully exploited are found in those regions. The principles discussed, however, have application outside of the tropics.

Within the tropics there are great differences in temperature and moisture regimes, which determine the natural plant ecology and the agricultural crop ecology, the ability of crops to survive, and their relative productivity. The agricultural systems of small farmers are determined, in the first place, by climatic conditions. Maximum, average, and minimum temperature during the crop season delimit the zones in which individual crops may be grown best. Further, within these temperature zones maximum, average, and minimum rainfall, as well as the seasonal distribution of rainfall and its predictability, determine where wheat, potatoes, sorghum, millet, corn, beans, rice, and other crops may be relied upon by small farmers with limited resources. The place for animals in the farming systems likewise depends upon the climatic limitations on feed, fodder, and crop by-products needed for their sustenance and productivity.

Soil and Water

Soils of the tropics vary greatly in natural fertility and in the physical qualities that determine water retention, internal drainage, and root penetration. The characteristics that are most important in determining fertility are geologic origin (volcanic, sedimentary, alluvial), chemical content of minerals and clays, weathering, erosion, and deposition of alluvial materials. The most important physical characteristics are particle distribution (sand, silt, and clay content), structure, and density. In addition, the terrain of agricultural land and that around it determines the natural drainage and the opportunities to modify the flow of water so as to irrigate. Soils with satisfactory physical properties may benefit from addition of fertilizers. Artificial changes in internal drainage may be made to mitigate or eliminate problems of salinity or flooding. Management practices of small farmers may reduce or enhance problems with erosion and deterioration of agricultural soils. These practices in the long run determine the net gains from expansion of land resources by clearing and reclamation or loss from reduction by man-made desertification. The agricultural systems of the small farmer not only provide for his living but also influence the future productivity of the land he works.

In areas where there are pronounced wet and dry periods in the year, land preparation (such as ditching and terracing,) and soil management (such as dry fallowing and contour plowing,) practices are used to adapt the farm site to the seasonal needs for drainage and water conservation. In dry climates, where irrigation is not available to supplement rainfall, crops having short growing seasons and deep rooting dominate the cropping patterns. Irrigation by diversion of river waters and use of reservoirs or by pumping subsoil water reserves permits the farming of suitable land in the driest regions; and by supplementing precipitation, it extends the cropping season and may permit intensive agriculture in regions with wet-dry climates. As population pressure or market demand for food and other farm products increase, the building of large irrigation facilities and the conservation of irrigation waters become matters of national concern. This need for water has justified enormous capital outlays and plans in the developing tropical countries of Asia, Africa, and the Americas. Although a high percentage of the expanded area under irrigation is devoted to large and medium-sized farms, an increasing proportion is set aside for small farmers. Where small farmers are involved, special provisions for cooperative water-users' associations and purchasing, marketing, and technical information organizations are essential components of successful irrigation developments.

Cropping Systems

The cropping systems of small farmers are comprised of the patterns of crops and their arrangement on individual fields, including their succession in an agricultural year and their rotation over several years, and the crop management practices followed—tillage, pest control, fertilization, and others. The array of crops from which the farmer selects is determined by their climatic requirements in relation to his site and by his economic needs and opportunities.

Intensity of land use varies greatly in different cropping systems. The most extensive and probably most ancient sedentary farming system is shifting cultivation. Because much time is required for regeneration of fertility of the land cleared for farming, ten to twenty times as much land is needed as that under cultivation at any time. The most land-and-labor-intensive system is permanent multiple cropping under irrigation. Between these degrees of intensity of land use are found the systems used in rain-fed areas with well-defined dry seasons that may or may not permit more than a single crop each year. Land productivity depends not only on the temperature and rainfall regimes and soil fertility but also on the ability of farmers to select cropping systems that exploit these resources to a maximum. The small farmers in an area, having limited capital and land resources, must choose cropping systems that maximize their contribution to family welfare at minimum risk. Any agricultural development program that seeks to improve the conditions of small farmers or encourage production of specific crops for sale can be successful only if it is based on full knowledge of the cropping systems used in each area and addresses the needs for economic or infrastructural inputs that will be required to improve those systems.

Farming Systems Including Animals

Animals are an integral part of the farming systems of the majority of small farmers with limited land and capital resources. Because animals depend on crops and crop residues for their subsistence and productivity, crop-animal mixed farming systems are based on the crops and cropping systems found in each small farming site. Animals have a broader range of climatic tolerance than crops and therefore may be important in areas having widely different climates. Thus cattle, swine, and poultry are found over a wide range of climates. But goats are concentrated in the drier climates and sheep in the cooler climates of the tropics. The importance of animals in the mixed farming systems of small farmers depends on the amount of land available, the cultural customs, the importance of animals as sources of power (transportation or draft), and the limitations imposed by endemic diseases. Animals not only contribute to family nutrition but also to economic goals that are specific to small farmers, such as negotiable savings to protect against unpredictable contingencies, fuel, and fertilizer. Thus, conceptually, the roles of animals in the farming systems of small farmers differ from region to region and are multiple—to satisfy needs for nutrition, power, fertilizer, fuel, economic security, and cultural obligations or ambitions. In order to understand how they fit into existing farming systems and how they may be exploited for improvement of family welfare, it is necessary to have basic information on the physical environments of specific small farmer sites and the cropping systems that are adapted to those sites, as well as on the socioeconomic factors affecting the small household.

Tools and Machines

The opportunities of the small farmer in tropical regions to exploit his land depend to a large extent on the power, tools, and machinery at his disposal. If he has to depend on his hands and those of his family for power and on simple hand tools for machinery, his opportunities are more limited than if he can supplement hand power with animal power and appropriate tools and machinery. If mechanical power provided by water or fossil fuel can be used to supplement or replace hand and animal power, the horizon of his opportunities expands. The choices are limited by his land, capital, and market resources and the benefit-cost ratio for each additional mechanical operation. To be economical for small farmers, large tractors and large machines powered by them must be managed cooperatively or must be available on a rental basis at the times in the agricultural year that fit the farmer's cropping schedule. Management of water and maintenance of canals for community irrigation systems must also be adjusted to the cropping programs of the small farm user. Specialized large harvesting, threshing, and drying facilities in small farm communties must meet the same conditions. The design of small power units and specialized land preparation, harvesting, and threshing equipment must be simple and low in cost and must provide for local repairs and maintenance. Adaptations of machinery manufactured for agricultural use in other regions have to be done locally in the context of the farming systems of the small farmers if their needs are to be satisfied. For this purpose, cooperation among the manufacturer, the dealer representing him, and local farm machinerey technicians may be essential.

Short- and Long-Term Effects

This chapter should not be concluded without brief reference to the short-term and long-term effects of farming systems on the physical environment in which they evolve. The immediate goals of small farmers—to provide food, comfort, security, and social status for the family—may be attained through farming systems that maintain, enhance, or destroy the physical resources necessary for the attainment of those goals. Only farming systems that maintain or enhance those resources can sustain permanent agriculture. Yet small farmers may use cropping patterns and grazing practices that cause irreversible losses of land as a result of erosion, salinization, or desertification. Since alternative practices that minimize or eliminate such losses may be available without significant reduction in the opportunities to attain the short-term goals, research on farming systems should focus on those alternatives, without losing sight of the possibilities for improved short-term productivity.

Enhancement of land productivity often depends on capital inputs for irrigation, drainage, or erosion control that require group action—either at the community or watershed level or at the regional or national level. National planning and international financing may be needed. Nevertheless, when such large programs are planned, it is essential that there be careful assessments of the ways they will affect existing farming systems. Research should be undertaken at the level of the small farmers, to develop and evaluate promising modifications in the farming systems, and development needs of the necessary infrastructure—both physical and institutional—at the community level should be identified and provided for.

Part III

Social Systems
from Farm Families
to National Programs

In Part I we reviewed the experience of international and national agricultural programs and of major regional R&D projects. We recognized that monocultural programs have made great contributions in creating new high-yielding plant varieties but that this research strategy has limited value for small farmers, who generally raise a variety of crops along with animals. This conflict led us to recognize the need to gear our thinking to farming systems. In Part II we presented an overview of the physical and biological elements that impose limitations and offer opportunities in the building of more effective systems for small farmers. In this analysis, we kept farmers in the background, mentioning them only generally to indicate how they are affected by the physical and biological elements of farming.

In Part III we bring our principal actors, the small farmers, to the foreground, considering their roles in building their farming systems. Just as it is useful to think of the physical and biological elements as systems, so also we need to think of human behavior as social systems and to consider how the social system may be integrated with the physical and biological systems (Chapter 10). In Chapter 11, we place farmers in the context of the local social system of family, household, and community.

Chapter 12, ''The Farming Systems Research Approach: The Rediscovery of Peasant Rationality,'' is designed to fill a gap between Parts I and II. Part I described how a recognition of problems with conventional R&D programs led to important insights regarding new directions for R&D but did not indicate how such insights might be converted into new ways of thinking about the research and development process or of new ways of organizing that process. Part II presented the basic physical and biological systems underlying farming systems but did not indicate how the physical and biological systems are to be integrated with the social system. The accomplishment of such integration requires that we change conventional ways of thinking about small farmers and the R&D process. On the basis of this new

127

conceptualization, we can then go on to devise social systems that are more effective in meeting the needs and interests of small farmers.

Chapter 12 describes how researchers went beyond the insights gained from studying the deficiencies of earlier programs to focus attention on what could be learned from small farmers and then begin to redirect research so as to involve small farmers actively in the process of experimentation and discovery. Here we pick up the story in Mexico where we left it with the Puebla project (Chapter 3) and link the emergence of new ways of thinking and acting in Latin America with similar trends in Asia and Africa. Chapter 13 examines the emergence of new R&D national programs in Guatemala and Honduras.

10

Systems Thinking
for Understanding
the Small Farm Enterprise

In Part III we focus our attention on social systems from family and household to community and to national programs. It would be more accurate to describe our focus as on political, economic, social, and psychological systems, but that terminology would be too cumbersome, so we will simply assume that the word ''social'' subsumes the other aspects.

In this chapter we seek to demonstrate the importance of developing a pattern of systems thinking as a means of dealing with the large number of variables operating in the process of agricultural and rural development.

Inadequacies of Simple Cause-Effect Models

No researcher really believes that a change in a single variable will yield a desired change in outcome. Yet the high degree of specialization in research tends to lead us to think that the variable on which we are concentrating is the key to agricultural development. Although the scientists who developed the high-yielding varieties of wheat and rice would not have made such a claim, the spectacular successes of these varieties tended to support the notion that a markedly improved technology would solve all problems. In fact, the spread of the high-yielding varieties has been extraordinarily rapid. For some parts of the countries where they were introduced the local adaptations of new varieties were so superior to existing genetic material that the change in one set of variables seemed to overcome all other variables, producing impressive increases in yields.

The perspective of hindsight should not diminish the importance of these great genetic advances, but it is now all too clear that basic problems of limited-resource farmers will not yield to any single technological fix. We have to recognize that no single factor represents *the* answer, whether it be high-yielding varieties, fertilizers, machines, better prices, cooperative organizations, or any other single variable.

We have to think in terms of systems—of the mutual dependence among a set of variables such that the change introduced into one leads to changes in others. What are these variables? How do they relate to each other?

The Problem of Complexity

Almost anything is somehow related to almost anything else, so that if we pursue the possible interrelations of variables as far as they lead us, we bog down in complex and confusing problems. We need to delimit the field in general form, work toward better approximations regarding relations among key variables, see what works in practice, and use empirical findings to arrive at improved formulations of our theoretical framework.

In this field, we should not think of a single system but rather of a number of systems that are related to each other in ways that we must discover. In the physical sciences and in some well-studied aspects of the social sciences, we want to specify the variables and develop precise measurements of relations among them. Since we are dealing with a broad range of variables across the animal, plant, and social sciences, it would be premature to seek such specificity. If we push too hard in that direction, we may fall into the trap of using only those variables which are easy to measure, neglecting others that would turn out to be more important.

What we are attempting here is not a tightly knit theoretical framework but rather a mapping of the theoretical domain. That is, we are undertaking to provide a general framework that will lead us toward further research that is of both practical and theoretical significance.

Natural and Physical Environment

What farmers are able to achieve in increasing their incomes and social and economic well-being will depend in large measure upon the natural and physical environment within which they live and work.

Some of these environmental elements are totally beyond the control of farmers. In some cases, however, creative adaptations are possible, and in other cases, farmers may themselves modify to some extent the physical infrastructure. Although major changes are generally beyond the reach of individuals or groups of farmers, in some cases public works projects carried out by government may introduce substantial changes.

Climate, for example, is clearly beyond the control of farmers, but resourceful adaptation to that climate may be possible. Altitude has a major impact upon temperature and therefore upon the potential for agricultural activities. In the tropics, some fruits such as the banana grow best at low altitudes, but white potatoes thrive at higher and cooler elevations, and cattle and sheep may graze at altitudes where it is difficult or impossible to grow most crops. But this does not mean that farmers are limited by the altitude where their household is located. For example, Steven B. Brush (1977) reports on one community in the Peruvian Andes which carried on its farming activities over a range of altitudes from 4,000 to 12,000 feet. In other words, with each general level of altitude presenting potentialities and limitations, farmers overcame the constraints of climate to some extent by spreading their activities over a wide range of altitudes.

The nature of the soil can also be regarded as a fixed condition but subject to modification within certain limits. Poor soil management practices result in depletion of fertility and may result in the washing away of much of the topsoil. On the other hand, good soil conservation practices preserve the fertility of the soil and enhance the farmer's ability to make a living from it.

Farmers can do nothing to affect the pattern of rainfall, but in some situations they can relieve water deficits by constructing tubewells to bring up subsurface water or by channeling river water into their fields. Here we need to distinguish between what can reasonably be done by small farmers themselves to improve their access to water and what can be done only, if at all, through projects financed and carried out by government. Tubewells may be beyond the reach of small farmers without outside financing and technical assistance. In many communities, the effectiveness of the irrigation system is determined by the farmers who manage that system, but whether they can achieve an adequate flow of water will depend not only upon whether there is a river nearby but also upon whether and to what extent they can use its water. Usually a government authority must manage the flow of water, not only by controlling the capacity of the river but also by balancing the needs and demands of various communities.

In many parts of the world, villagers have provided labor to construct roads that link them with markets, but whether they can build roads without substantial outside help will depend not only upon the terrain to be traversed but also upon the rainfall. For example, in many areas where there is a heavy rainy season some roads may become impassable during this period, and a major public works project would be necessary to provide all-weather roads.

At this point, we do not wish to deal with how and to what extent farmers may modify the physical environment. We should consider the existing environment in which people live at the time of our observations as a baseline against which we need to assess potentialities and limitations. Only when we have systematic knowledge of the potentialities and limitations of the environment will it be profitable to consider changes that might be introduced. When we can distinguish between changes that may be accomplished by the villagers themselves and changes that require outside assistance, we will need to consider the socioeconomic infrastructure and the socioeconomic system of farmer-household-community.

The Local Social System

Here we come to our primary focus of attention: the small farmer. In order to place the farmer in a realistic context, we need to link him or her with the family farm enterprise and the community. In considering larger commercial farmers, it may be practical to carry out the analysis in traditional economic terms, assessing the land, labor, and capital involved in the enterprise, focusing upon the individual farmer. Since small farmers may hire little or no outside labor, depending upon family for farm labor and for organizing the consumption activities of the household, we must see the farmer in the context of family and household. The household not only provides labor for the farm. In many cases, members work part or full time off of the family farm, and their income is important to the household economy. The potential for using additional family farm labor also has to be seen in the context of competing outside earnings opportunities.

In many parts of the world, farmers draw upon labor beyond the family, through traditional forms of labor exchange or through hiring their fellow villagers. Thus the community provides an important base for the human resources necessary beyond the individual farm. The farmer may be able to tap these resources through personal relationships. In other words, within this local social system, we are considering only those elements that are subject to mobilization by the personal efforts of the farmer. We think it is important to conceptualize this level separately from the commercial and politico-administrative infrastructure.

For those activities potentially under control of the farmers themselves, we consider management opportunities for, and constraints upon, improving the standard of living of the family. Here we think of management of soil, water, fertilization, and pest, weed, and disease control.

If the land is served by irrigation, then necessarily the farmer must be involved with others in the management of this system, which leads us to examine the functioning of an important organization, which is part of the local social system. In this same category, what farmers can do collectively to create and manage efficient associations or cooperatives to provide them with economies of scale in purchasing and marketing will strengthen the ability of the individual farmer to enhance the economic well-being of the farm family. Such organizations also have important administrative and political significance because, as we shall see in Chapter 14, they may become elements in linking individual farmers with agriculture-related government organizations. They may also provide an important political base for small farmers, thus enabling them to influence state policies more effectively.

Commercial and Politico-Administrative Infrastructure

Organizations such as agencies and institutions for agricultural research, extension, the system of credit through banks, and markets are largely under outside control. In most countries, the market organized by private entrepreneurs is predominant, but in many countries, a government purchasing organization may buy some staple products at fixed prices and thus have some influence upon marketing. Policies and programs of government agencies such as agricultural price policy are of great importance. In some countries government imposes price policies for the benefit of the more politically active urban consumers to the disadvantage of farmers.

The agencies seeking to help the small farmers may provide assistance, effectively or ineffectively, in various forms. Research and extension may provide valuable information and ideas. The state or private organizations may help the farmer get better access to the inputs needed and may provide assistance in marketing. Banking institutions may provide agricultural credit at low to exorbitant rates of interest, delivered in timely fashion or too late to do much good.

The state may play an important role in the building of public works, such as a dam to store water for more effective irrigation. The state may also provide storage facilities to help the farmers withhold their produce from market when prices are lowest.

All of these forms of assistance can be useful to small farmers, but more often than not we find that a single item of assistance, such as production innovation or a loan, will not enable the farmer to make a major change in the economic well-being of his family because what is needed is a coordinated set of resources. One of the major problems of agricultural development is the lack of coordination among government agencies, each of which carries out its own specialized functions with little regard to how those functions fit into the total activities and needs of small farmers. Thus any system that can achieve coordination among government agencies will provide major improvements in services to farmers—a topic to which we will devote attention later.

Figure 10.1. Guide to systems thinking

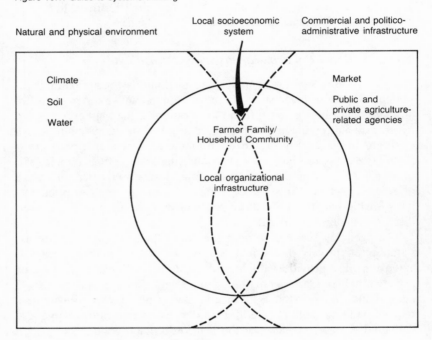

Visualizing the Interrelations of Systems

What we have said may be presented in diagrammatic form, as Figure 10.1 indicates. There we see the local social system embedded in two other systems: the natural and physical environment and the commerical and politico-administrative infrastructure. The encompassing systems overlap to a minor degree to indicate that actions in the commercial and politico-administrative system can introduce changes in the natural and physical environment. That is, climate cannot be changed except through long-run shifts in patterns of cultivation and deforestation or reforestation. But the availability of water can be affected in major ways by government public works projects or by government promotion of tubewells.

We see the natural and physical environment offering opportunities to farmers and also imposing constraints upon them. Thinking about the local social system shows us how farmers seek to deal with these opportunities and constraints, individually and collectively. The commercial and politico-administrative infrastructure provides its own set of opportunities and constraints for local farm families.

In examining the problems of any farm family, we need to place it in relation to the local social system, the natural and physical environment, and the commercial and politico-administrative infrastructure. The aim of this book is to aid students and practitioners to think more effectively about the complex interrelations involved in this approach to systems analysis for agricultural research and rural development.

We should recognize the limitations as well as the potentials of systems thinking in this field. We do not see systems thinking leading directly to propositions useful for policy decisions, but we believe this approach can provide a useful way of thinking about complex phenomena and can protect us against misleading oversimplifications. Even if the planner cannot measure certain variables, it may be important to recognize their existence and seek to estimate their potential impact upon any development plan.

11

Farmer, Family, Household, and Community

Before we can make sense out of the human side of farming, we need to liberate ourselves from semantic problems that affect the way we think about agricultural and rural development. Whether the farm is large or small, we are accustomed to thinking and talking of "the farmer." The problem with this way of thinking is that it leads us to visualize a single individual or an aggregation of individuals. Furthermore, although "the farmer" does not specify sex, we tend to visualize a man.

These semantic traps lead us to think of the motivation of individual males and of the results of their actions. There are, of course, two problems with that way of thinking. First, a farmer is hardly ever an individual who lives alone and works his farm by himself. Generally he is a member of a family and lives in a home, which must be seen as a household providing the base for the producing and consuming operations of those who live there. We must therefore visualize farmers as playing various roles in a complex social system revolving around the household.

Second, we are likely to think of the male farmer as head of the household, the individual who directs agricultural work, with his wife bearing his children, preparing food for the family, keeping the house in order, and taking care of the children. This picture is highly misleading. In up to one-third of households, the woman is the head and manages the agricultural work, either because she is a widow, because her husband is away earning money elsewhere, or because in her particular culture, most of the farm work falls to women (Boulding, 1977). In many cases, even when women are not heavily involved in agricultural work, they play a major role in marketing the produce of the farm.

In peasant communities, it is usually assumed that the male is head of the household and is dominant in family decision making. Although this may often be the case, we must look beyond public appearances, as a Peruvian anthropologist discovered in his experiment in applied anthropology in an Indian community in the Andes (Núñez del Prado, 1975). Seeking to involve the community of Kuyo Chico in democratic processes, Oscar Núñez del Prado began holding periodic meetings with the adults for discussion and decision making on possible improvement projects. Adults of both sexes were invited to these meetings, but at first those attending were predominantly males, and the few women who attended remained in the background and did not participate in the discussion.

Only after several abortive attempts to carry out projects that appeared to have the unanimous support of the men did the anthropologist recognize the defect in his procedures. After the meeting, the wives would confront their husbands at home and ask why they had agreed to such a foolish project. Of course, the men did not report back that the wives had talked them out of their decisions, but neither did they go ahead with implementation. Until he was able to involve the women in group decision making in the public meetings on a more equal basis, Núñez del Prado followed a policy of avoiding reaching a decision in a single meeting on any idea for action that he or anyone else presented. He arranged to have the idea thoroughly discussed in the first meeting and asked the people to come to the next meeting ready to make a decision. In this way, the men were able to sound out their wives on the proposed plan without having to acknowledge publicly that the woman played a major role in deciding what got done in the community.[1]

Recognizing the important role played by women in agricultural work and also in household and farming decision making has important implications for the agricultural research and development process. In most countries around the world, rural extension services are built upon the assumption that the male directs the agricultural work and does that work himself, while the female confines her work to the traditional functions of home economics. Therefore, males are trained and hired to provide agricultural extension to rural males, and, when women are employed in extension, their responsibilities are generally directed toward the traditional tasks of homemaking. We have found case after case in which a male agricultural extension officer seeks to move the rural male toward the adoption of a new agricultural practice without success because, among other reasons, it is the female who does the agricultural work in questions. Even when the male extension agent recognizes that the woman is primarily involved in the activity he is

[1]Jon Swanson and Mary Hebert (personal communication) report a similar two-step process in their Cornell rural participation research in Yemen. For other such examples, see Staudt (1975) for Kenya.

seeking to guide, that does not solve the problem. In many societies, it is not thought fitting for a male who is not a member of the family to spend much time with a woman, and it may often be difficult for the extension agent to arrange for proper chaperonage whenever he tries to communicate with the female on the farm.

In many peasant communities, even very young children contribute labor to the family farm. In the Peruvian highlands, children as young as six may tend animals out in the grazing area as well as providing some of the care for small animals and fowl near the home. Generally, boys work under the direction of their fathers and girls with their mothers.

As children grow older, their labor becomes more valuable to the farm, and yet various forces prevent the full use of that labor. Children in school can devote less time to the farm, and as children advance further in school they may have to leave the community to continue their education. Peasants may well doubt the value of education for improving the efficiency of their farm, but they are often well aware that opportunities to earn money off the farm are likely to be strongly affected by their children's educational credentials. In this situation, parents have to balance present needs against future gains.

Especially for adult males, off-farm income may be necessary to maintain the household. Without this supplementary income, it might be impossible for a family to survive on the farm. In fact, we may find families with off-farm income that are better off financially than families with more land but no off-farm income.

Davydd J. Greenwood presents a picture of the complex socioeconomic system of the peasant household:

> The calculus of maximization on the peasant farm must include considerations of subsistence requirements, "rents" (Wolf, 1966), the farming cycle, and markets for both subsistence and cash crops, not to mention various domestic group obligations. Dowries may be provided to daughters to ensure them a proper marriage. Boys may be apprenticed or otherwise given a start in life to permit them a self-sufficient adulthood. Care for the aged and sick must be provided out of farm resources. Apart from this there are community obligations to be satisfied, by providing work, paying levies, and otherwise supporting common activities. Ceremonial obligations must also be met for the good of the family and the community as a whole.   [Greenwood, 1973:51]

We might add the need to provide education for children to enable them to fit into the off-farm economy and then to contribute some of their earnings to the household.

Land as the Base of the Household Economy

In addition to land, the resource base for the farmer includes household tools, machines, physical structures, and money to buy inputs. But since no farming is possible without access to land, we may consider land as the fundamental resource for the farm family.

Volumes have been written about the causes and consequences of different sizes of landholdings, of various systems of land tenure, and of programs for "land reform." These topics are too complex to be fully covered here, but we can suggest some points to keep in mind in thinking about the relations of farmers to their land.

In the first place, we must recognize that there are millions of farmers who do not depend entirely on what they grow for support of their households. In many cases, the household income drawn from crops and animals is supplemented by earnings of family members away from home as farm laborers or urban workers. Development planners must not assume that, if the farm is too small to provide full subsistence for the family, the only solution is to encourage them to leave the land and seek urban employment. An improvement that brings the family closer to self-sufficiency in farming may be well worth the effort of all concerned. On the other hand, if the proposed improvement requires more labor, farmers and extension agents need to consider whether that additional labor can be supplied by those currently working on the farm. If the additional labor requirement can be met only by a family member giving up paid work off the farm, these costs must be weighed against the expected benefits from the improvement.

Other things being equal, we can assume that the farmer with more land will be better off than the farmer with less. Apart from the abilities of the farmers, conditions vary so enormously from country to country and region to region that the generalization has little practical value. In many parts of the world, population increase has resulted in division of land from one generation to the next, so that the holdings become smaller and smaller. In general, this process will lead to the impoverishment of the community, but this may not always be the case. Consider, for example, the village of Huayopampa on the western slope of the Andes in Peru. In 1927, 51 percent of the families had holdings under one hectare. By 1966, the number of families farming less than one hectare had increased to 74 percent. In that period, however, Huayopampa had shifted from the traditional crops of maize and potatoes to much more commercially profitable fruit crops and had become one of the most affluent villages in the highlands (Whyte and Alberti, 1976).

We can expect on the average that the owner of land will be better off than the tenant, but much depends upon the nature of the rental contract, since in some cases tenants who have access to relatively large and fertile expanses of land may be in a better economic position than owners who farm smaller and less favored holdings. Whether the tenant is better off than the sharecropper will depend upon the nature of the rental contract and the shareholding arrangement. We naturally assume that the owner of a farm must be better off than the sharecropper on the same size and quality of farm. This may be true on the average, but not in every case. If the sharecropper has been entirely dependent upon the landlord for seeds, fertilizer, tools, and equipment, or has been dependent on him for credit to purchase necessary inputs and tools, then a government decree giving the sharecropper title to the land will not improve his situation unless the government or some other agency steps in to help him get the inputs and other resources he needs under conditions more favorable than those he had with the former landlord.

Comparisons within and among Peasant Villages

If every farmer managed his farm differently from every other in his area, it would be impossible for the professional to discover a common baseline representing the indigenous farming system. In fact, if the area selected for study has common ecological conditions and its farmers are similar in culture and control over resources, we can expect to find farm management practices and strategies conforming to a common pattern (Gostyla and Whyte, 1980). Having established the general nature of that pattern, it is important to discover variations from one segment of the population to another. In comparing one village with another, even in the same region, we may find wide differences within the same general pattern or even farming systems that are distinctly different from each other. Students of development need to be concerned both with uniformities and with variations in farming systems in a given ecological area and within the larger region.

Let us illustrate these points with cases from two parts of the developing world.

Latin American Cases. In the highlands of Guatemala, maize is the basic subsistence crop. The socioeconomic unit of the Instituto de Ciencia y Tecnología Agrícolas (ICTA, the Guatemalan agricultural research institute) has found that what farmers do and wish to do regarding their farming practices depends to a large extent upon the adequacy of their maize harvest for family subsistence (Ruano, 1980; Gostyla and Whyte, 1980).

In this area, the small holders may be divided into three strata. At the bottom are those who cannot raise enough maize to cover the subsistence needs of their families throughout the year. They are concerned with gaining off-farm income, but on the farm their primary interest is in increasing the maize harvest, so as to reach the subsistence level.

At the middle level are farms that provide enough maize to feed the family and perhaps a small surplus to be sold in the market. These families are less concerned about raising the maize yield than about other plants that might give them a cash income, in addition to home consumption. For this group ICTA has pursued a strategy that both maintains the existing maize yield and offers farmers cash income for growing wheat.

At the top stratum are farmers who can readily cover their subsistence needs in maize, with the possibility of a moderate surplus to sell in the market. These farmers are more secure in their basic subsistence crop, so they are more inclined than the others to experiment with other crops. For them, ICTA has developed a strategy of interplanting wheat and cabbage or other vegetables with maize.

Even between communities of small farmers in the same ecological area and growing the same major crops, we may find significant differences in wealth based upon differences in the work activities within families. For example, as we were driving through an area between Chimaltenango and Guatemala City, Peter Hildebrand informed us that "just a little farther along this road, you will begin to see pickup trucks by peoples' homes." Indeed, the pickups appeared in fair numbers.

A superficial look at the terrain before and after the appearance of the pickups revealed no ecological differences to account for the differences in wealth. Furthermore, it was evident that maize was the major crop in both communities. Hildebrand explained that, in the poorer community, the women of the household spent much of their time in the traditional craft of weaving. In the more affluent community, for reasons as yet unknown to the researchers, the women did not weave and instead devoted themselves to truck gardening for the Guatemala City market. They needed the pickups to get the produce to market. Raising vegetables was more lucrative than weaving and so they could afford the vehicles.

Even within a general agricultural pattern in a given community, there may be substantial income differences. Examples are Huayopampa and Pacaraos, two villages on the western slopes of the Peruvian Andes, Huayopampa being at an elevation of about 6,000 feet and Pacaraos being somewhat over 10,000 (Whyte and Alberti, 1976). We found marked differences in average monthly income between the two villages and also marked differences in the distribution of income, as shown in the Figure 11.1. Although there was little difference in average farm size between the two communities, Huayopampa was far more affluent that Pacaraos and showed a far more equitable distribution of income within the community. The range in earnings per family from top to bottom levels was about 18 to 1 in Pacaraos compared to about 4 to 1 in Huayopampa. Furthermore, the largest stratum (close to 35 percent) in Huayopampa was in the middle. As the distribution suggests, Huayopampa had a large and affluent middle class, whereas Pacaraos had a very small affluent group, far separated from the poor and very poor.

The differences in average earnings between the two villages can be explained only by the contrast between the farming systems. After 1948, Huayopampa shifted from traditional crops to commercial fruit growing, mainly citrus fruits. These more valuable crops set off such a boom in the local economy that at the time of our studies in the mid-1960s, half of the families in Huayopampa also owned homes in Lima. The colder climate of Pacaraos did not permit a shift in crops planted.

Although the prosperity of Huayopampa was general among all families who were *comuneros* (landowners and community members), there were marked differences in the way those in different strata worked the land and used labor.

Five families at the top, with an average of 1.64 hectares of irrigated land, hired labor and owned their own machines. Nine families at the second level, with an average of 1.15 hectares, hired labor and rented machines. Fifty-seven families in the third stratum, with an average of 1 hectare, made little use of hired labor or machines and depended upon traditional forms of reciprocal labor among relatives and close friends. In the fourth stratum, 28 families, with an average of 0.76 hectares, supplemented their income from fruit with part-time labor for more affluent comuneros. At the bottom, with an average of 0.54 hectares, 52 families concentrated on their own properties and did not enter into reciprocal exchanges or seek part-time employment.

Below these five strata were two small groups of families: landless farm laborers and sheep and cattle herders who worked in the pastures far above Huayopampa. Being migrants into the community, being generally

Figure 11.1. Monthly income distribution in Huayopampa and Pacaraos, Peru

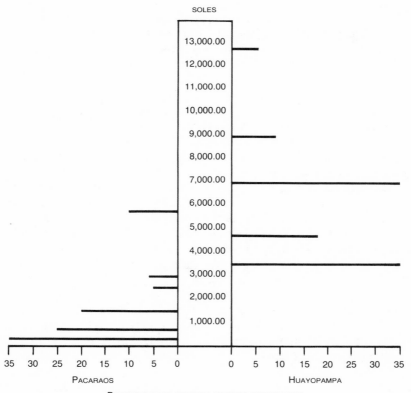

PERCENTAGE OF FAMILIES AT EACH INCOME LEVEL

Note: For Huayopampa, although the principal income is from fruit, additional income from other sources is estimated from field reports. At the time of these estimates (1967), the sol was valued at 26.80 to the United States dollar.
Source: Whyte and Alberti (1976:166).

illiterate, and not being comuneros, they played no active role in local affairs.

The top social stratum earned a higher income from fruit growing than did the lower social strata entirely because of differences in land hold-ings. The difference in fruit-growing income per hectare between the top and bottom strata we estimated at an insignificant 28 soles per month. And since the families in the bottom stratum did not hire labor or use machines, their net income per hectare from fruit growing was markedly higher than the top families' net income per hectare from fruit growing. The greater affluence at the top levels depended not only upon more land but also upon professional and commercial incomes. Since families in the third stratum, depending upon reciprocal labor, not only saved on labor costs compared with those above them but also had more sheep and cattle than any other stratum, their net family income was close to that of stratum 2. [Whyte and Alberti, 1976:175]

Asian Cases. In her studies of Nepal, Jacqueline Ashby (1980) found systematic differences among three villages located in the same area but at different altitudes. She divided her total sample into four farm types based on differences in relation to the market, in outside earnings, and in the use of labor.

If we review Table 11.1 from the bottom stratum up, we note that those with the least land are not the poorest. The subsistence farmers have slightly more land than the part-time farmers, who have a substan-tially higher income. The part-timers have so little land that they could not possibly hope to support themselves from their land alone and thus send some of their members off the farm to bring in 85 percent of the family income. The subsistence farmers are much more heavily depen-dent upon their agricultural produce, which brings in 68 percent of their total income. They supplement their farm income by a relatively small proportion of off-farm earnings (13 percent) compared with the part-timers. They are more dependent upon getting outside income through working on lands of more prosperous farmers, gaining almost a fifth of their income from this source compared to the part-timers who gain only 1.5 percent of income from farm labor. Even the subsistence and part-time farmers find it necessary to hire labor at times of peak work requirements for over sixty man-days per year.

In this system, the subsistence farmers are particularly disadvan-taged. They need all of the labor the family can provide and so find it exceedingly difficult to send children away to school with the hope of gaining off-farm income in the future. The part-time farmers have as many workers available as the subsistence farmers, and, with almost

Table 11.1. Productivity by farm type in Nepal

Farm type	Number of households	Sales[a]		Cash earnings[b]				Income[a]		Land	Labor[c]		
		Percent crop production sold	Percent livestock production sold	Average cash earnings (rupees)	Farm sales as percent of cash	Nonfarm earnings percent of cash	Agricultural wages/as percent/of cash	Average gross income per capita (rupees)	Nonfarm earnings as percent of total	Average area cultivated (hectare)	Average number days hired	Average number family workers	Average family workers per hectare
Subsistence	156	10	12	396	68	13	19	540	1	0.57	66	2.5	4.4
Part-time	67	6	38	2,880	13	85	1.5	900	39	0.41	67	2.5	6.2
Small commercial	48	27	71	4,620	96	4	—[d]	1,320	2	0.67	145	2.9	4.4
Large commercial	61	40	17	6,624	70	29	—[d]	1,608	13	1.80	467	3.6	2.0

Source: Adapted from Ashby (1980: 119)
[a] Sales and total production valued at farm gate price for each area.
[b] Farm sales valued at price reported for sales by individual respondents.
[c] Male and female, aged 15–55 years for family labor.
[d] Less than 1 percent.
72 rupees = 1 U.S. dollar

one-quarter less land to farm, they are more easily able to spare children for schooling.

Only at the level of the small commercial farmers do we find households selling a substantial proportion of their crops (27 percent) in the market, and this figure increases to 40 percent among the large commercial farmers. Selling of livestock is important for part-time farmers, who sell 38 percent of what they produce, but still more important for small commercial farmers, who sell 71 percent.

Students working on a Cornell Agricultural Economics Department project have provided important information on the household economies of relatively poor farmers in Asia (Sisler and Colman, 1979). They were particularly concerned with the adoption of modern improvements in farming such as the new seed varieties, inorganic fertilizer, herbicides, and insecticides. They also examined the implementation of physical changes such as improved water control and mechanization through the use of tractors, mechanical threshers, and pump sets.

In general, they found little difference in the rate of adoption of modern varieties, inorganic fertilizer, and insecticides. Farmers with less than a hectare seemed to adopt these improvements as rapidly as those with one to three hectares or more than three hectares. On the other hand, they were less able or willing than larger farmers to adopt tractors, mechanical threshers, and pump sets and to use herbicides.

Figures on yields, earnings, and costs provide a simple explanation for these differences in the adoption of different elements of modern technology. Researchers found no difference in yields per hectare through the use of tractors or mechanical threshers. These items, which simply replaced labor, were of no practical value to the small farmer, who is more likely than the larger operator to have sufficient family labor available to meet most or all of the needs of his farm. The comparison between the adoption of insecticides and the nonadoption of herbicides by the small farmers has the same explanation. Insecticides had a markedly favorable impact upon yields, whereas weeds could be eliminated through manual labor, and the farmer did not have to buy the chemicals and sprayers needed for herbicides.

Farmers who cannot afford mechanization may find that the increased amount of labor they devote to their fields actually results in higher yields than those achieved by the large farmers. For example, Gillian Hart found in Indonesia that the smallest farmers (averaging .12 hectare) used 76 percent more labor per hectare and were rewarded with 59 percent higher yields than those of the largest operators on an average farm size of 3.15 hectares (Sisler and Colman, 1979).

These findings give an encouraging picture of the efficiency of small farmers. Studies in various parts of the world have generally found smaller farmers getting higher yields per hectare than larger farmers. Apparently, the small farmer provides more "tender loving care" to his property than do the larger operators. Economies of "detail" more than offset economies of scale. This suggests questioning the common assumption that it is lack of capital that holds back the small farmer. If by capital we mean such large investments as tractors, mechanical threshers, and other expensive machines, it is clear that these expenditures would not pay off for the small farmer even if he had the money to invest. If use of such equipment is needed to make possible the planting of a second crop or to avoid harvest losses, it must be made available by cooperative ownership or hire. On the other hand, the less expensive purchases of inputs such as modern plant varieties, inorganic fertilizer, and insecticides may be critically important, so that the small farmer who is unable to purchase these items finds himself at a great disadvantage.

Which variety of rice was planted in what amounts was not a matter of peasant traditionalism or resistance to change in the villages studied by Douglas Pachico in Nepal (Sisler and Colman, 1979). Rather, it was a matter of achieving a delicate balance between the needs and resources of the family. The farmers were choosing among two modern varieties, Taichin, a nitrogen-responsive dwarf variety, Pokhareli, a comparatively high-yielding Napalese variety, and Thapachina, which had been the traditionally grown local variety. Taichin is the highest yielding of the three, but it is more difficult and time-consuming to thresh than Pokhareli. Taichin's advantage of a shorter growing season is offset by local judgment that it is somewhat inferior in taste and cooking quality. Pokhareli requires more transplanting labor than Taichin, and the Pokhareli plants are frequently bound together before harvest to prevent lodging (bending down of the stalks). This practice increases labor requirements before and during the harvest period. Thapachina, the local variety, has markedly lower yields than Pokhareli, but it also has a much shorter growing season and excellent cooking qualities.

Pachico described the factors going into varietal selection:

(1) the higher yielding Taichin is preferred by small farmers operating close to subsistence, but with adequate family labor to cover the harvest peak;

(2) larger farmers who must hire labor, react to the cost and difficulty
of obtaining harvest labor by growing a relatively high proportion
of Pokhareli, which has a lower harvest labor requirement than
Taichin; and

(3) larger farmers combine a higher proportion of Thapachina with
the other two varieties because its early maturation spreads the
harvest labor peak, and it provides fresh rice at an earlier date for
festivals. [Sisler and Colman, 1979:15]

This example shows that selection of varieties is not a simple question
of one being better than another in all respects. Each variety must fit
into the household economy and work plans.

Tractors and mechanical threshers contribute no increase in yield for
their owners, but they do have a major impact upon the local economy
and social system. Both in the Philippines and Indonesia, researchers
found that mechanization by the larger farmers was breaking down the
traditional custom in which smaller farmers and landless laborers had
an open invitation to work at harvest time for the large farmer in
exchange for a fixed percentage of what they harvested. The larger
farmers found that the machines enabled them to carry out the harvest
with much less labor than previously required, and it was thus to their
advantage to minimize the labor they used and to pay cash for that
labor. The authors suggest that when farms are increasingly mechanized, the large landowners need the small farmers and the landless
much less than before and have the mechanical power and greater
income to enable them to expand their holdings further. Small farmers,
having lost the opportunity to supplement their income by working for
the large farmer, may be forced to abandon the land to the large
operators. Use of capital instead of labor is more profitable for the large
farmer but less efficient for the entire society in a poor country where
capital is more scarce than labor.

Modernization and Labor Exchange

Technological progress also tends to disrupt the traditional pattern of
social relations. In many parts of the world, small farmers exchange
labor among family and friends. The farmer's household is required to
provide food and drink to those who come to work on the family land.
These customs not only serve an economic purpose but also strengthen
social ties among the members of the exchange network.

Although this custom has a natural appeal to observers of peasant
life, it has disadvantages in achieving efficient management and the
adoption of new methods (Erasmus, 1961). The system works best

when all of those engaged in the work bee are doing the same familiar things that they do on their own farms. As any farmer differentiates himself from the customary ways, he faces increasing problems in getting help. Furthermore, the work-ceremonial system does not lend itself to good discipline. Since the work activities are carried out in a spirit of free exchange and voluntarism, the farmer can hardly prod his friends and relatives when they seem to be putting more spirit into eating and drinking and sociability than into labor. When the farmer breaks out of this social custom and hires labor, he is in a much stronger position to demand compliance with his directions.

External Relations

In the past it was customary to consider the peasant village as being more or less outside of the market economy. This view is misleading. The problem is not that the peasant village is outside of the economy but rather that it is plugged in in a way that is disadvantageous to the peasants. Therefore, in assessing the position of the family or household, we must see it not only in the context of the community but also in the context of external relations of the household and community to the market and the politico-administrative structure.

We have already noted the importance of the labor market for the farm household, observing that in many cases household income from farming is supplemented by income gained from urban work or labor on other people's farms. Let us now turn to the produce market, which is the major means whereby the household turns its farm work into cash income.

The ability of the farm household to profit from the market depends in large measure upon its access to the market. Access is not simply a matter of distance but depends upon the presence or absence of a road and the quality of that road, whether it is usable all year or unusable in times of heavy rains, availability and cost of means of transport, and so on. When the farmers of Huayopampa could get their produce to market only on muleback or on their own backs, they were limited to the traditional crops. Only when a road gave them reasonably easy access to urban markets were they able to shift to perishable crops and increase incomes enormously.

When extreme inequality in the distribution of wealth and power exists, not even a good road can guarantee ready access to market. For example, the Indian community of Kuyo Chico had a reasonably good road passing near the village and leading into the mestizo-dominated market town of Pisac (Núñez del Prado, 1975). When the Indians could get their produce to the market, they were able to sell at competitive prices. Up until the late 1950s, however, on market days they were often intercepted on the road by *alcanzadores* who relieved them of their produce at prices about half of those prevailing in the Pisac market. In this era, physical resistance was not practical because the *alcanzadores* were backed by the political and police powers of Pisac. Furthermore, the merchants of Pisac had agreed that no one would sell kerosene to the Indians for cash. The Indians could get the kerosene only through barter, offering eggs or other produce. This system enabled the merchants to make a double profit by overpricing the kerosene and underpricing the produce. This exploitation prevailed until a community development project undermined the political power of the mestizo authorities and also established in Kuyo Chico a cooperative for buying and selling kerosene.

If the peasant has more to sell than he himself can get to market, he must depend upon middlemen who come to the village by truck to buy the produce. This method would not be disadvantageous to the peasants if there were active competition among a number of middlemen, but in developing countries, this is seldom the case. There are so few truck owners that the peasants may be able to deal only with a single operator or, when there are two or three, they may together fix prices. Furthermore, in peasant society the relation between the peasant farmer and the middleman is likely to be more than a pure commercial transaction. Since the truck owner travels frequently between the villages and the city, he is in a position to carry messages and even packages both ways. His ability to do favors for the peasants helps him to establish a patron-client relationship, thus further limiting the peasant's ability to operate in a competitive market. The extreme of this patron-client relationship was found some time ago among peasants living and farming along the tributaries of the Amazon River in the Peruvian jungle. There the river merchants provided their own mobile market, buying peasant produce and selling to the peasants. In much of this region there are no roads, so that the river is the only transportation. The peasant's family may have its own small canoe or other nonmechanical craft, but it may have to depend upon the river merchant to get a family member to a clinic or hospital in case of a medical emergency.

Governments may seek to break the dependence on middlemen through establishing an agency that guarantees minimum prices and provides an assured market for a limited number of crops commonly grown by the peasants. This system can offer the peasants a significant price advantage, but government payment is likely to be slow and undependable. For example, in the potato-raising community of

Huasahuasi in the Peruvian Highlands, the heavy rains generally come around harvest time. It was difficult for the small farmers, who lived far from the road, to get their potatoes out for roadside pickup. Then if they did not sell the potatoes promptly they could lose their crop in the continuing downpour. When the rains were most heavy, the road to market was almost impassable, so officials would keep the government trucks in the garage. Under these circumstances, an independent truck owner who would risk the trip could buy the potatoes at a small fraction of their worth in the market. Even when the government driver did get his truck to Huasahuasi, he was empowered to pay with paper certifying the amount of potatoes he had purchased at a price to be determined by the quality of the load. Quality determination and pricing would be done by professionals at regional headquarters. Usually the farmer would have to wait three months for payment, and then generally he got less than he expected because of discounts for poor quality. On the other hand, whatever price the middleman paid, he gave cash on the spot. Since he was in business for himself, he was free to make his own calculations of risk and prices and pay at the time of purchase (Whyte, 1977).

Peasants escape dependence on middlemen or on an undependable government agency through owning their own vehicles, but few households in poor communities can afford such capital investments. An obvious alternative is cooperative ownership in which a number of families band together, pool their resources, operate the truck, maintain it, and use it to carry their produce to market and return with supplies. Such organization can offer important advantages of scale of operations, strengthening the bargaining power of members in the market. It is also a potential base for the exercise of political power, since organized peasants can exercise more influence than unorganized peasants. Beyond the obvious problems of repairing and maintaining a complex new technology in a peasant village, however, the success of the cooperative depends upon the ability of the members and their leaders to develop and manage a large organization. We will examine further the problems and potentials of cooperatives in Chapter 14.

Paradoxical as it may seem, in some cases peasants may gain advantages through bartering their produce rather than entering into the cash market. For example a potato farmer in a highland village in Peru began his marketing trip with one sack of potatoes and ended a series of transactions with eight sacks (Burchard, 1974). After the end of the harvest each year, he would put a sack of potatoes on his back and set out on foot toward the lowlands east of the Andes. Some days later he would arrive at a village that raised no potatoes but had an ample supply of coca leaves. He would exchange the sack of potatoes for a sack of coca leaves and walk back to his village, where he would exchange the sack of coca leaves for eight sacks of potatoes. It might be argued that there was little profit in this transaction if we considered the value of labor expended. But during this period there were no remunerative activities carried out on the farm, so the opportunity costs of foregone income were extremely low. Besides, the farmer followed this pattern year after year, visiting with friends and relatives along his route, so he gained in social relations to compensate for his labor.

Conclusion

Although we naturally tend to speak of *the farmer*, we need to think of a household unit rather than a single individual. We also need to think of that family and household within the context of the community and the market beyond that community.

We need to visualize the household as a small social system made up of male and female adults and generally also of children whose roles in the farming and household activities vary with the crops grown, the size of the farm, the number of agricultural cycles in a year, and so on. We have noted that women often are important in decision making in farming and even more generally in farm labor. When it comes to labor, we need to think not only of what the household applies to its own farm and the labor it hires but also of the labor performed by household members on other farms in the area and off the farms in the urban areas, with such earnings returning to support the farm household.

We recognize that labor is acquired for the peasant farm not only by cash payments but often through exchanges that not only get the work done but reinforce social ties among relatives and friends. This labor exchange tends to become inefficient as farmers differentiate their activities, requiring new tasks and more diligent performance than is generally available through the traditional system.

When we consider the effect of "modernization" upon poor farmers, we must distinguish among types of innovations that fall under the general heading. The studies reported for Asia suggest that new high-yielding plant varieties, chemical fertilizer, and insecticides are adopted in many situations as rapidly by the poor farmers—assuming they have access to such improvements—as by the wealthier farmers. Mechanization that reduces the need for labor offers no advantage to the small operators in the cases we have examined, although it may be financially advantageous to the big farmers, enabling them to reduce their needs for hired labor. The advantages to the large operators often come at the expense of the poor farmers. Losing their opportunity to add to their income through part-time work with the large operators, they may be unable to make a living by farming and thus have to abandon their land. As these abandoned farms are picked up by the large operators, the process of modernization through mechanization increases the differences in income from the top to the bottom of rural society and may lead to explosive social tensions.

We have noted that one cannot understand the situation of the small farmer without examining his relations to the market. Here we have examined a wide range of factors affecting small farmer–market relations.

Finally, we have noted that the individual small farmer is at a great disadvantage in dealing with the market and the politico-administrative structure. Since economies of scale can make a difference both economically and politically, it is useful to consider organizations of farmers and rural people generally as important parts of the local infrastructure. In Part IV we will examine more fully organizational elements of this local infrastructure, as we consider also the relationship between the local and the national infrastructure provided largely by government.

The Farming Systems Research Approach: The Rediscovery of Peasant Rationality

Before we move on to new ways of thinking and acting, it may be useful to go back to the roots of the agricultural sciences, as reflected in the following statement, made by a scientist more than seven decades ago:

> We had long desired to stand face to face with Chinese and Japanese farmers . . . to walk through their fields and to learn by seeing some of their methods, appliances and practices which centuries of stress and experience have led them to adopt. We desired to learn how it is possible after twenty and perhaps thirty or even forty centuries, for their soils to be made to produce sufficiently for the maintenance of such dense populations as are living now in these . . . countries. We have now had this opportunity and almost every day we were instructed in the ways and extent to which these nations for centuries have been conserving and utilizing their natural resources; we were surprised at the magnitude of the returns they are getting from their fields.   [King, 1911:2]

This desire to learn from farmers—and even from small farmers—was not limited to scientists who traveled abroad. In the late nineteeth and early twentieth centuries, many agricultural scientists in the United States and Europe spent much of their time out on farmers' fields, observing and interviewing farmers. In this early period, scientists recognized that their contributions would have to be based upon an intimate knowledge of farming systems actually in use. Furthermore, since most of them had grown up on farms, it was natural for them to value what they could learn from experienced farmers.

In our admiration for the feats of biological and chemical science in recent decades, we are inclined to overlook the fact that it was not until the 1930s that the knowledge, experience, and genetic materials accumulated in the two previous generations of agricultural sciences brought about large increases in yield per unit of land. To be sure, it was no mean feat to hold yields steady over decades in the face of crop

154

pests, soil erosion, weeds, and other problems, and the new science contributed to this achievement. In the United States up to that time the major advances in agriculture had been derived largely from increasing the productivity of labor through mechanization, which enabled farmers to expand the land area they cultivated without corresponding changes in labor input.

As agricultural scientists came to concentrate their activities increasingly in laboratories and on experiment stations, on-farm research and inquiry tended to receive less emphasis. Agricultural scientists in the United States did not lose contact with farmers but tended to interact particularly with the larger and more successful farmers.

By the 1970s, observers were coming to recognize that the benefits of the Green Revolution had been very unevenly distributed and that the majority of small farmers in developing countries, cultivating rainfed areas, had received relatively little benefit. In some cases, smaller farmers had even lost ground as the larger and more favored farmers had prospered.

In the 1950s and 1960s, movement of "surplus" rural population to the cities in developing countries was welcomed by some economists. Industrial expansion was expected to provide urban employment for the migrants, while the reduction of the rural population would lead to larger and therefore more efficient farming operations.

By the 1970s it was apparent that the course of change had departed drastically from the theory. Heavy urban migration indeed continued, but in general, urban employment was expanding far too slowly to absorb the influx of potential workers. Furthermore, rural birth rates remained high enough to counterbalance the outmigration and generally maintained the preexisting level of rural population. Whether a trend toward larger farms with less labor input would have increased total agricultural output—a hypothesis which most empirical studies seem to contradict—became a moot question as the predicted trend failed to materialize.

In many developing countries, large farms had been devoted traditionally to export crops, with smaller farms primarily producing food for the rural and urban population. As urban population expanded and domestic food production failed to keep pace, food imports created increasingly serious foreign exchange problems in many countries. These economic problems tended to push the policy makers toward a reexamination of their agricultural development strategies.

As researchers came to recognize that the benefits of the new technologies were not trickling down to the small farmers, they focused attention upon presumed barriers to the transfer of technology. In past decades, behavioral scientists had contributed to a monumental misunderstanding of this problem—an erroneous diagnosis that is implicit in the very concept of "transfer of technology." Perhaps because they were trying to convince plant scientists of the value of social research, behavioral scientists simply assumed that the recommendations the agricultural researcher or extension agent gave to peasant farmers were bound to be economically beneficial to them and their families. Since, more often than not, these farmers failed to adopt the recommendations, it was assumed that the problem must be cultural: the peasant farmer was locked into the traditional system of beliefs and practices. Therefore, the problem was one of "overcoming resistance to change."

At this point, we should recognize differences in interpretation among the various social science disciplines. As late as the 1960s, belief in the myth of the passive peasant was particularly prevalent among behavioral scientists (social anthropologists, social psychologists, and sociologists). Development economists were more inclined to assume peasant rationality (see particularly Schultz, 1964). On the other hand, influenced by their social class position, many administrators and program planners tended to assume that peasants did not know what was good for them and therefore had to be induced to accept the "improvements" that their superiors had planned for them.

As we abandon what we now call the myth of the passive peasant, we view the small farmer as a rational being who seeks to balance gains and losses and to minimize risks. We do not assume that he always makes the correct decisions in his own interests in adjusting to the particular conditions he faces. But we do assume that twenty to forty years or more of experience in farming in a given area has given the farmer an intimate, practical knowledge of behavior of plants and animals in that area under varying conditions and that the agricultural scientist needs to gain access to the information and ideas of the small farmer if he is to be able to make any useful contribution to that farmer and his farm. We are now coming to recognize that the knowledge of small farmers is more than the accumulation of experience, handed down from generation to generation. Indeed, social scientists have discovered farmers carrying out their own indigenous experiments.

Consider, for example, the following case from Nigeria:

> In one case, people experimented with cassava when it was first introduced. As cassava can be poisonous, it was important to establish the conditions in which it could safely be eaten. The procedure adopted was to feed it first to goats and dogs. In another case, a scientist believed he

had made a breakthrough when he found a way of breeding yams from seed, propagation normally being vegetative. A farmer was casually encountered, however, who had not only himself succeeded in doing this, but had also discovered that whereas the first generation tubers were abnormally small, the second and subsequent generations were of normal size. The scientist reportedly exclaimed, "Thank God these farmers don't write scientific papers." It was also noted, in support of the prevalence of experimentation by farmers, that there is a Yoruba word for "experiment." [Howes and Chambers, 1979:6]

The emerging new research strategy involved three principal elements: a shift in emphasis away from monocultural or single crop or animal product research toward research on farming systems especially adapted to the needs and interests of small farmers; a shift in emphasis away from the experiment station and toward on-farm research with active participation of small farmers; and realization that the planning, conduct, and evaluation of this research requires interdisciplinary activity, not only within groups of biologists and social scientists but also between them.

Agricultural scientists came to recognize that farming in the tropics, though more complex, offered more opportunities than in temperate zones where there is generally only one growing season (except for winter wheat). Thus where the climate is warm enough and where rainfall is adequate year around, or where irrigation can compensate for dry seasons, agricultural activities can include two to four successive crops per year. The various cropping systems that have developed under tropical conditions were discussed in Chapter 6.

New Directions for Research in Asia and Africa

A pioneer in this new line of research was Richard Bradfield, who had been a key figure in the 1940s in launching the Rockefeller Foundation's Office of Special Studies, which led to the development of high-yielding varieties of wheat in Mexico and to the creation of CIMMYT. In the 1960s, when he became particularly concerned about the productivity of small farms (especially under tropical conditions), Bradfield was working with the International Rice Research Institute in the Philippines. Inspired by what he had observed on small farms in Taiwan, Bradfield devised a system characterized by an extraordinarily intensive use of land through intercropping, relay planting, and sequencing of planting so as to get three or four full growing seasons within a given year.

The Bradfield system required an extraordinarily high level of farm management skills and very large expenditures for inputs and use of machines, as well as abundant irrigation water and so was not directly applicable to the conditions of small farmers. But the enormous yields achieved by Bradfield impressed many scientists with the potential for improving the income of small farmers through more intensive use of their land.

Richard Harwood, also working in Asia, took the essential step of moving from the Bradfield method of working at the experiment station into the farmers' fields and began developing participatory experiments with farmers that have been reported in his important book (Harwood, 1979). In Africa, M. P. Collinson (1972) and David Norman took the lead in investigating mixed cropping under indigenous conditions (Norman, 1973, 1980). Others began to give increasing attention to the indigenous cropping systems of Africa.

An example of the reorientation of research can be seen in the fact that researchers began to question the value of plowing tropical soils in the way traditionally done in temperate climates. Tropical soils often give better results if not disturbed by plowing. This discovery has led to a growing number of experiments in ''minimum tillage''—a fancy new name for a very old principle. Ironically, scientists have been finding that, in some conditions, the ''primitive'' digging stick is a more useful tool than anything provided by modern technology. (Now many farmers even in temperate zones are experimenting with minimum tillage.)

Cropping Systems Research in Latin America

During the same time period, important research on indigenous farming systems was also going on in Latin America. As early as 1971–1972, the Tropical Agronomy Center for Research and Training (CATIE) had begun experimentation to develop its own adaptations of the IRRI multiple-cropping strategy. Beyond operating its experiment station at Turrialba in Costa Rica, CATIE maintains agricultural scientists in other Latin American countries, where they work with national programs, expecially providing technical assistance on intercropping research. Although the work of Bradfield and others at IRRI was known to some agricultural scientists in Latin American at the time they began work on indigenous cropping systems, it seems likely the major impetus there came from the discovery in the Puebla project that farmers interplanting maize and beans were getting much more value from their fields than those who practiced the monocultural system originally recommended by the professionals. By all accounts, the man most influential in gaining acceptance of this reinterpretation was

Leobardo Jimenez, who had been the first field director of Puebla and later became dean of the Colegio de Postgraduados at Chapingo and subsequently deputy director of the National Agricultural Extension Service in Mexico.

Antonio Turrent, who was head of agronomic research on the Puebla project, in 1968 was one of the first scientists to grasp the significance of intercropping for small farmers and to pursue systematic research in this field. In one experiment, Turrent compared yields of maize and beans separately with yields of the interplanted crops, using the same patterns of fertilization in each case. He found that beans yielded just as much interplanted as when planted alone, and the maize yield in association with beans was approximately 70 percent of that achieved in monocultural planting. At prices then prevailing in the area for the inputs used and for the produce sold, Turrent found that the maize-bean association yielded 54 percent greater net income than maize alone and 113 percent greater income than beans alone (Turrent, 1978).

In Guatemala, experiments showed that the maize-bean association not only yielded the same amount of beans as monocultural planting of that crop but also yielded somewhat more maize than when maize was planted alone. The explanation for this difference in the results of Turrent's and Donald Kass's experiments is found in the pattern of fertilizer use in the two cases. True to the traditions of agronomic research, Turrent made his comparisons between monocultural and intercropping patterns using exactly the same amount of fertilizer in each experiment. Kass followed the customary practices of Guatemalan small farmers, who were not accustomed to using fertilizer on maize, which they grew primarily for home consumption. They did customarily use fertilizer on beans, some of which they expected to sell in the market. When maize and beans were interplanted, the maize picked up some of the fertilizer laid down for the beans and thus naturally yielded better than when maize was planted alone, without fertilizer (Kass, personal communication, 1978).

Turrent was particularly impressed with a farming system developed by the small farmers of Oaxaca, which made extraordinarily efficient use of scarce water. The system had not gradually evolved over centuries of trial and error, as might be assumed in the case of the maize-polebean association. The castor bean had been introduced in Oaxaca during World War II in response to the demand for its oil, which was especially valuable in certain industrial operations. After 1945 the government no longer promoted the growth of the castor bean plant, but many of the small farmers found that they still had an attractive market.

The advantage of the castor bean plant in Oaxaca is that it has exceptionally deep roots. In a maize–castor bean association, the two species are planted at the same time just at the start of the rainy season, which provides light rains over a period of six months. The corn is harvested at the end of the rainy season, and the castor plant is left standing. By this time its roots have extended deeply enough to continue to absorb moisture, and the plant continues to grow for another six months. At the end of that time, farmers cut down the plants, which are quite large, gather the beans for sale, use the castor stalks for firewood, and use the leaves as cattle fodder.

Turrent's experiments, again with identical treatment of fertilizer, show that maize planted alone yields somewhat more than when planted in the most dense pattern of the castor bean, but the most profitable interplanting combination yields 2 to 6.7 times as much income, depending upon the market price of castor beans. This income advantage of the combination does not take into account the value of the leaves as cattle fodder or of the plant stalk as firewood in an area that has little fuel available.

Stillman Bradfield commented upon the Oaxaca interplanting system:

> I did a small study of the maize–castor bean association and found that there are a number of other advantages. . . . The castor bean can be harvested over a long period of time, then stored for a long time in the house, without deterioration, allowing children, old folks, etc., to crack open the pods to get the seeds out as needed for cash. It apparently does not spoil so it permits them a marvelous spread of farm labor throughout the year, utilizes household labor, and generates cash whenever needed.   [Personal communication, 1980]

Bradfield added that the Mexican government has been contemplating establishing a guaranteed price of at least 9 pesos per kilo for castor beans. Turrent noted that the advantage of the maize-bean association over maize alone ranged widely because the price of castor beans fluctuated from 2 pesos to 10 pesos per kilo. If the government fixes a castor bean price close to the top of this range, the small farmers of Oaxaca stand to gain substantial benefits.

The new orientation involved a shift in emphasis in experimentation away from the experiment station and onto farmers' fields. By 1979, official reports of the Instituto Nacional de Investigaciones Agrarias (INIA) showed that 56 percent of all experiments were carried out on the fields of small farmers.

As experiments on farmers' fields proceeded, several Mexican scientists recognized that the full potential of the integration of indigenous agriculture with the modern agricultural sciences could not be gained simply through multiplying the number of on-farm experiments. To go beyond the simple determination of what works and what does not, scientists needed to discover the underlying logic of the farmers' systems of cropping developed under diverse ecological, cultural, and economic conditions. Ethnobotanist Efraín Hernandez X., of the Colegio de Postgraduados has been directing an ambitious study of indigenous cropping systems in three areas of Mexico. In this project, supported by the Mexican government, Hernandez has worked with six anthropologists, four biologists, and three agronomists. This selection of personnel illustrates another important lesson of the Puebla project: the study of indigenous cropping systems necessarily is an interdisciplinary enterprise.

The areas selected for study represent a wide range of agricultural conditions and systems in Mexico. Southern Yucatan is characterized by slash-and-burn or swidden agriculture. Oaxaca is a very dry and poor agricultural area, but settled agriculture is practiced. El Bajio, to the north of Mexico City, is regarded as the bread basket of the capital district. That area is favored by rich soil, and much of the land is irrigated so that farmers can produce under more favorable conditions.

During the same period in Colombia, ICA, the national agricultural research organization, was becoming increasingly involved in interdisciplinary on-farm research. The close ties between ICA leadership and the professional staff of the Caqueza project and the movement of Caqueza staff members and students with project field research experience into positions in ICA further enabled ICA to build upon Caqueza experience.

By the late 1970s scientists were ready to draw general conclusions regarding the advantages of intercropping. As Turrent writes, "the patterns of cultivation that form parts of the system of peasant agriculture are rational and, while there exists an ample opportunity to improve them, as a general rule the productivity of the land under these patterns of cultivation is potentially greater than that which is achieved with monocultural systems" (Turrent,1978). Other researchers elsewhere have found additional potential advantages of intercropping under certain conditions: better soil conservation in areas of heavy tropical rainfall and improvement in pest, disease, and weed control (see, for example, Innes, 1980).

Although intercropping was increasingly emphasized in research programs during the 1970s, the integration of animals into farming systems research tended to lag behind because in most situations there has been much less contact between animal and plant scientists than among specialists within various plant science disciplines. The emphasis on cropping systems research, however, has highlighted the need to bring animals into the emerging new framework. Robert E. McDowell (personal communication) has pointed out that certain new high-yielding varieties of grains are not suitable for small farms that include cattle. For example, development of a variety of maize with tougher stalks to resist the corn borer and shorter stalks to concentrate more of the plant's energy into production of the ears of corn would be disadvantageous to the farmer who also raises cattle. The "improved" variety offers less volume of fodder and a higher lignin content, thus reducing its nutritional value for animals. Similarly, the shorter-stalk rice varieties developed by IRRI to maximize grain yield are too high in lignin content to be suitable fodder for buffalo or cattle.

As the importance of integrating animals into farming systems research is being increasingly recognized, we now see ILCA, the International Livestock Center for Africa, doing research on animals in relation to cropping systems. IITA, the International Institute of Tropical Agriculture, having encountered soil depletion problems in its cropping research, is now integrating animals into its programs.

Intercropping, of course, provides a major barrier to mechanization. The farmer may freely use the tractor for plowing and harrowing before he does any planting, but he cannot take the tractor through a densely interplanted field after the plants have begun to grow. The severity of this limitation depends upon the total farming system being practiced and upon the household economy of the farm family, a subject we will consider later. And, as noted in Chapter 11, mechanization generally does not provide increased yields to small farmers.

Conclusion

The new approach to on-farm research is well stated by one of the pioneers in this area:

> The planning process involves the scientist with the farmer in deciding what modifications and innovations to try. Each brings to the planning process his own perspective and his own wisdom. The farmer contributes his intimate, often tacit, understanding of his own situation and the factors that influence his productivity. The scientist has the objective information derived from his measurements and observations, plus a familiarity with alternative production technologies from other areas. The scientist and the farmer collaborate on planning and implementing changes, and the results are measured against mutually agreed-upon

goals. The careful documentation of their experience with new technologies and systems in well-defined environments makes it possible to extrapolate their results to other, similar situations in any part of the world.

This approach depends to a great extent on teamwork among scientists whose disciplines are highly specialized and insular and who are unaccustomed to working together on common problems. The process proposed in this book requires agronomists to work with crop and soil scientists, animal specialists, agricultural economists, nutritionists, and educators. Interdisciplinary collaboration is crucial to the process, and the team includes a coordinator whose special function is to bring the disparate insights and skills of the various scientific specialists into focus on the problem of increasing the small farmer's production.   [Harwood, 1979:7–8]

Harwood recognizes that it is not sufficient simply to locate research on farmers' fields:

The farmer's actual participation in the planning, execution, and evaluation of research should be clearly distinguished from mere research in farmers' fields initiated and controlled completely by scientists. The latter approach simply provides a test of technological components in various actual farm environments. The results may be valuable to the scientists, but they do not show how well the new technology performs under the farmer's management, nor how it integrates into his farming system. And they do not encourage the adoption of successful innovations by the farmer-participant.

It is crucial that the research organization appreciate the value of joint farmer-scientist planning, testing, and evaluation of technological changes. The farmer's criticism or rejection of the researcher's favorite methods or new varieties is often difficult for the researcher to accept. It involves both his personal and his professional pride. But if the farmer's opinion is ignored, discounted, or even ridiculed, the fragile connection between farmer and researchers on which this entire system depends will be broken.   [Harwood, 1979:40–41]

13

Toward New Systems of Agricultural Research and Development

Chapter 12 described the evolution of thinking about the integration of small farmers into programs for agricultural research and development. This chapter focuses on ways in which these ideas are being implemented in emerging organizational models. Such models must be designed to solve two problems:

(1) To devise a system of on-farm research built upon the active participation of small farmers. It is not enough to demonstrate that, in one experimental project on fields of small farmers, encouraging results were achieved. To convert isolated cases into a systematic regional or national program, planners need to specify a set of procedures to be carried out jointly by professionals and farmers and then go on to develop an organization that can carry them out.

(2) To integrate the on-farm research program into the already established national programs of experiment stations, extension, credit, and marketing. It is not enough simply to add on-farm research to the other preexisting programs. Unless on-farm research can be linked effectively with the well established components of the national program, the new element will come to be seen as a fad to be abandoned when other new ideas come along.

Here we concentrate on the shaping of new organizational models in Guatemala and Honduras. Guatemala's Instituto de Ciencia y Tecnología Agrícolas (ICTA) provides one of the most clear-cut examples of a new organizational model for agricultural research. ICTA has had problems in integrating this new research model with the other components of the national agricultural program. Honduras adopted essential elements of the ICTA research model and at the same time achieved a more effective integration of research with extension and other agricultural programs. Thus the Honduras case will show how one nation learned from another and then improved upon the organizational model it adopted.

ICTA in Guatemala[1]

Guatemala entered the 1970s with a history of about fifty years of agricultural research and development but without an integrated national program. Activities were fragmented among agencies with little coordination. As the 1960s came to an end, government leaders and agricultural professionals were concerned about the rising tide of food imports. This drain on the national economy focused high-level attention upon the need to increase agricultural production. An important feature of the five-year development plan for agriculture (1971–1975) was the government's creation of ICTA as a relatively autonomous research institute but at the outset operating within the large existing organization devoted primarily to extension, the Dirección General de Servicios Agricolas (DIGESA).

The principal planners and organizers of ICTA were agronomists Mario A. Martinez, then vice-minister of agriculture, and Astolfo Fumigalli, the new director of research for ICTA. They agreed upon a two-point program. If the research institute was to fulfill its mission, it must be separated from DIGESA. (Although they were probably correct in assuming that ICTA required substantial autonomy in order to develop an innovative research program, its complete separation from DIGESA led to problems which became apparent in later stages.) They also saw the importance of linking ICTA with international centers and with foundations supporting agricultural research and development.

The Guatemalan planners were successful in interesting the Rockefeller Foundation in their new program. The foundation supported a planning conference and then agreed to assign Robert Waugh, an animal scientist with rich experience in developing countries, to ICTA as consultant. ICTA began operations in early 1973 as an autonomous unit, independent of DIGESA, under the joint direction of Martinez and Fumigalli. At the outset, the planners had a clear conception of ICTA's objectives but no more than a general sense of direction. The way this sense of direction was translated into methodology, organizational structure, and social processes shaped the institute's future program.

The early plans for ICTA called for on-farm research focusing particularly on the needs and interests of small farmers. An increasing

[1]For a detailed presentation of the evolution of ICTA, see Gostyla and Whyte (1980). ICTA is one of several national research institutes that are innovating along the same lines. We concentrate on ICTA because we are able to provide a detailed account of its evolution, based largely on field work.

amount of research was carried out beyond the experiment station, but no clear methodology had been developed. It was the creation and evolution of a socioeconomic unit in ICTA that stimulated the development of what we now consider the distinctive ICTA organizational model for agricultural research.

The socioeconomic unit was established in 1975 under the direction of agricultural economist Peter Hildebrand, who came to Guatemala after several years experience in El Salvador, where he had developed a highly innovative and productive research program based upon systems of intercropping and close collaboration with peasant farmers. ICTA first conceived the role of the socioeconomic unit as to evaluate the impact of current research and development activities. Therefore, one of its first assignments was to evaluate complaints of farmers in one area against the imposition of certain recommendations as a condition for their receiving bank loans from DIGESA and BANDESA, the agricultural development bank. To receive loans in La Maquina farmers had to participate in a supervised credit program and use the credit to buy substantial amounts of fertilizer recommended by DIGESA and BANDESA. Farmers had been complaining that the fertilizer had a negligible effect upon yields in their particular area.

Observations by members of a technology testing unit in La Maquina had indicated that the farmers might well be right, but it remained for the socioeconomic unit to analyze the costs and yields of cooperating farmers to substantiate that conclusion. After some study, ICTA reported that in La Maquina, fertilizer had such a slight effect upon yields that the expenditure was not worth the cost. This report led to an important change in the policies of DIGESA and BANDESA: the farmers were no longer required to use any of their loan for the purchase of fertilizer. (To be sure, some extension agents still continued to advise its use, almost as an act of faith, but it was no longer compulsory.)

Although this case demonstrated the potential usefulness of socioeconomic research, it could hardly serve as a model for the development of a program to fit into the activities of the other units in the Ministry of Agriculture. Evaluative research would inevitably place the socioeconomic unit in the position of criticizing the work of professionals in other disciplines, other parts of ICTA, or other agencies of the ministry.

Hildebrand and his associates therefore sought to involve themselves in the early stages of the research process instead of simply coming in to evaluate the work done by others. Finding the necessary starting point was not easy. Hildebrand was impressed with the enormous gap between conditions on experiment stations and those outside on the peasant farms. Invariably, the stations had been laid out and developed in the most favorable conditions for obtaining maximum farm yields. The land was relatively flat and fertile and amply supplied with water. Farm machinery was available, and there was an ample supply of the inputs scientists considered necessary to obtain maximum yields. Not far from the experiment stations, the small farmers were struggling to eke out an existence on hillsides, on rocky terrain of low fertility; they were using bullocks for plowing and were able to afford far fewer inputs than recommended by plant scientists, based on their experiment station program.

Hildebrand sought to persuade one experiment station director to substitute bullocks for tractors and move most of his experimental program off the station and onto the hillsides typical of peasant farming. The proposal provoked an indignant rejection. In fact, in the early months, members of the socioeconomic unit were seen by the station plant scientists as unreasonable and aggressive cranks, and the unit was unable to get any cooperation from the established experimental program. Finding itself blocked in fitting its program into the established structures, the unit won top-level approval to develop its own methods of on-farm research.

The social scientists began with a study to delimit an area where the farming system practiced by small farmers was relatively homogeneous. The purpose of this survey was to make sure that successful experiments would provide conclusions fairly applicable throughout the area. At the same time, to get systematic information on indigenous farming systems, the socioeconomic unit developed a program of *registros,* simple farm management records, to be filled out daily by the farmer or a member of his family, recording the amount and type of labor, the tools and power sources used, the amounts of fertilizer, pesticides, or other inputs applied, and so on. Members of the socioeconomic unit worked with the farmers to develop a balance between the researcher's desire to have a highly detailed quantitative record of farming practices and expenditures and the small farmers' need to work with a system that was simple enough for them to understand and might prove to be more helpful than burdensome.

Not having access to land on experiment stations, the socioeconomic unit rented small plots from local farmers and paid the farmer whose land it rented for the labor he provided in the experimental process. The aim was not to use the small farmer as a hired hand but to involve him as a consultant (*asesor*) and participant in the research-planning process. The unit proposed to try only those innovations that its farmer-consultants considered reasonable and promising. The rationale for this decision was that any innovation that seemed impractical to local farmers was not likely to gain acceptance.

The strategy was to start with minor changes, and especially changes that required little or no additional expenditure for inputs compared to the farmer's traditional practices. If a modest experiment yielded concrete benefits, the farmer would be encouraged to undertake further and more far-reaching changes.

The first on-farm experiments were carried out under the direction and control of professionals. Any innovation that did not work out at this stage was referred back to the plant scientists on experiment stations and in the regional organizations for advice and further study. The innovations that yielded good results moved into the second stage of farmer field trials. At this stage the socioeconomic unit gave up control and shifted into the role of consultant and observer. Farmers now tried out on their own fields, with their own money and their own unpaid labor, the innovation they had tested earlier. Innovations that did not work at this stage were referred back to the plant science professionals for advice and further study. Those that did work were assumed to be ready for diffusion and general adoption throughout the farming area. In this stage, the socioeconomic unit had become involved in the diffusion process, which is normally thought to be the jurisdiction of the extension service.

To avoid innovations requiring inputs beyond the means of small farmers, the unit concentrated upon developing new patterns of interplanting and use of space. For example, in one location in eastern Guatemala, it studied farmers' traditional *milpa* system of maize, sorghum, and bean interplanting, discovering that the most serious limits on production were quantity of bean seed and labor during the planting season, which in this area was limited to the dry period of two or three weeks following the first rains and before the onset of the heavy and continuing rains. Land was not a limiting factor; constraints of labor and available bean seed prevented farmers from fully using the land they owned.

The farmers' traditional patterns, used as a control for all experiments, and the new patterns are illustrated in Figure 13.1. In the traditional pattern (A), farmers planted maize and sorghum in alternate rows about .63 m apart, with beans intercropped at random between the rows. The socioeconomic unit changed the traditional pattern by introducing alternative population and spacing patterns. Maize and sorghum were planted in double rows at a distance of .315 m. Maize seedlings were placed diagonally between sorghum seedlings in the parallel row,

Figure 13.1. Patterns of maize-sorghum-bean associations used traditionally (A) and modified to provide less competition for light and greater plant population (B)

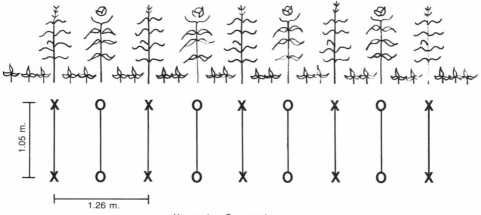

X = maize, O = sorghum
(beans intercropped at random)

A. Traditional system

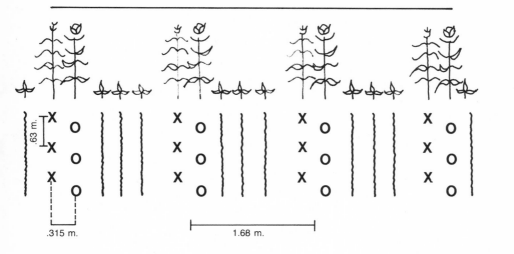

X = maize, O = sorghum

B. ICTA socioeconomic system

in a chainlike effect, as illustrated in Figure 13.1B. The diagonal planting pattern allowed for adequate sunlight to penetrate through to both crops. Within rows, populations of both maize and sorghum were increased relative to the traditional system. The distance between the centers of the double rows was 1.68 m. Ample space was left to plant three rows of beans at a population density of 48 percent of the traditional system. The double row arrangement allowed more open space for beans, at the same time that it increased maize and sorghum populations.

In the traditional system, beans consumed the majority of farmers' planting time. In the new system, with the reduced bean population, farmers were allowed additional time in planting so they could extend the area under cultivation onto land traditionally left fallow. The new system allowed the farmer to plant 40 percent more land than before, with the same amount of planting labor and with less bean seed, yet bean production was held stable. From the additional land under cultivation, the farmer produced 75 percent more maize, 40 percent more sorghum, and 33 percent more income. Productivity of both bean seed and labor at planting was therefore increased, relative to the traditional system. The socioeconomic unit claimed that the new system offered these advantages with minimal risks to the farmer. There was no additional requirement for fertilizer or pesticide beyond what he had been using traditionally.

In another area of Guatemala, studies by the unit revealed that land was the most severely limiting factor and capital was also relatively scarce. It identified three strata of farmers and then devised a production system suitable to the financial capacity of each stratum. Each system was designed to increase productivity per unit of land.

In Guatemala, maize is the most important staple in the diet, and the farmer's first concern is to raise enough to feed his family. For the farmer in the lowest stratum, first priority was to achieve self-sufficiency in the production of maize, while having little or no investment capacity. The socioeconomic unit devised a system of production that simply replaced single rows of maize with double rows (in a zigzag pattern to allow for air circulation and ample sunlight). The system called for 50 percent more maize seeds on the same area. Trial results indicated that the system could produce 45 percent more maize, allowing farmers in the lowest stratum to reach self-sufficiency.

A farmer in the middle stratum usually was able to produce enough to fill his family's subsistence needs, and he might sell small amounts to provide modest capital for new ventures. Here the unit devised a system involving double rows of maize 2 m apart, or twice the space allowed in the traditional system. In this open space, farmers planted wheat. Maize production dropped slightly with this system, but farmers were rewarded with a crop of wheat that sold at an attractive support price established by the government.

A farmer in the top stratum had no trouble producing enough maize for family consumption and had higher levels of capital to invest in improving his production system. Here the socioeconomic unit experimented with the maize-wheat combination just described but also introduced cabbages within the wheat stand. Trials showed that almost 14,000 cabbages per hectare could be grown without appreciably decreasing wheat production.

While refining its methodology for on-farm experiments and field trials, the socioeconomic unit was also improving and speeding up its system for making baseline studies of farming systems in an area. In the past, the traditional style of doing social science research has been an obstacle to its integration within agricultural R&D programs. Such research takes so much time that conclusions are reached after planners wish to act. Furthermore, the action implications of such research are often difficult for the administrator to discern.

The socioeconomic unit changed this pattern. At first, it conducted exploratory research to determine the major agronomic and socioeconomic features of the area; results were made available immediately to other programs for planning purposes. The unit then proceeded to study the area more in depth, with final recommendations ready for publication a year later. Eventually, ICTA leaders decided that the longer-range study was too costly and not necessary for planning. The unit now concentrates on improving its capacity to carry out exploratory research that can be quickly applied by agronomists. At present, the unit does a reconnaissance of an area in one or two weeks' time. It has become more familiar with the general characteristics of systems used by farmers; this knowledge facilitates the survey process by directing interviewers' attention toward key aspects of the farmers' practices.

Agronomists from the regional team are now participating with the unit in the survey process. Researchers are organized in pairs, consisting of one natural scientist and one social scientist; members are rotated daily within pairs to control against interviewing bias. At the end of each day, the members meet to discuss their work. They try to identify common patterns in their findings and fill in weak spots by following up upon particular themes the next day. This activity provides valuable cross-fertilization of information between disciplines.

The reports that the unit produces from these activities are available almost immediately upon completion of the field work, and they are written in clear and precise form that natural scientists can understand. The participation of agronomists in the surveys has helped the unit direct its research to areas that are of technical concern to the rest of the institute, and at the same time it has put technical people more in touch with farmers' problems. These reports are becoming increasingly useful in regional planning processes.

So far we have described the evolution of the socioeconomic unit methodology and given some indications of its contribution to ICTA but have not focused upon the problems of integrating it into a research organization dominated by plant scientists and organized according to two structural principles: crop specialization and regional decentralization. Directors of crop improvement programs (maize, beans, sorghum, and so on) worked in the central office, guiding experimentation in their particular crops in the seven regions. The headquarters of each region were located on an experiment station, now renamed production center. It was the responsibility of the regional director to coordinate the activities of his various specialists who were developing research on cropping systems and conducting experiments on farms as well as at the production center. How to coordinate these new lines of research had to be discovered in practice.

At first, many regional professionals reacted negatively to the onfarm experiments carried out by social scientists. Hildebrand describes these early reactions:

The year was very dry and had two prolonged periods without any rain. . . . Visitors . . . were surprised if not appalled to see field trials under such conditions, and the crops demonstrated the extreme stress under which they were growing. But it was also evident that these conditions were the reality under which the farmers of the Ladera lived and produced. Aside from the comments that it looked just like a trial being run by social scientists and that it was a good thing it was well off the road, the most usual comment was that it was obviously not worthwhile to work under these conditions because nothing could be accomplished. [Hildebrand, 1978]

Unless they are deliberately testing the impact of particular insects or plant diseases, the training and experience of plant scientists naturally lead them to take pride in growing plants that look healthy and promise high yields. They were shocked at the shoddy appearance of the socioeconomic units trial plots. This reaction indicates why it was necessary to allow the unit to go its own way at first, so as to develop its methodology, and also indicates the difficulties it was to experience in seeking to integrate its program into regional structures.

Leaders of the central organization facilitated the communication of the unit's findings and procedures in central and regional meetings, and occasionally a regional director began to take an interest in its farming system surveys and farm records and was planning to test further in order to make sure he came out with the best combination.

In 1977 the ICTA administration decided that a member of the socioeconomic unit should be assigned to each regional production center. This was an important step toward integration, but problems still remained. In the first place, the socioeconomic unit was not fully enough staffed to be able to place university graduates in the regional centers, and therefore had to send out *peritos* (agricultural high school graduates), whose lower status placed them at a disadvantage in dealing with the university graduate *ingenieros agrónomos*. Being especially qualified through experience working closely with farmers, the *peritos* were expected to guide the professionals in developing farmer records and in carrying out on-farm intercropping experiments according to the socioeconomic unit's methodology, but university graduates did not respond readily to the guidance of high school graduates.

In this case, more than a status problem was involved. In general, the crop improvement program heads in the central office and the regional directors had not made adjustments in work loads to allow for collecting farmers' records and for expansion of the on-farm research within work already in progress. The professionals naturally tended to do their accustomed work first and delay the new tasks being brought to them by the *peritos*.

In the course of a year, ICTA leaders noted a marked improvement in regional response to these new responsibilities initiated by the socioeconomic unit, as professionals from its central office were able to spend more time in the field with regional directors and their staffs. These professionals were able to help regional *peritos* to fit their work into the established programs and to demonstrate how socioeconomic unit projects could strengthen these programs. The new methodology for field surveys was a major influence in strengthening the unit's relations with plant and soil scientists in the regions. Since the surveys were carried out by pairs consisting of a social scientist and an agronomist from the regional organization, the success of this program helped representatives of the two units to appreciate the values each unit was bringing to the joint effort. By the 1979 planning meetings, regional directors were generally reporting that farmer records and area agrosocioeconomic surveys had become basic elements in their programs.

The improved internal integration of ICTA, however, did nothing to improve its relationship to DIGESA, which continued to operate under a model imported much earlier from the United States. The two agencies hold incompatible assumptions regarding the nature of small farmers. ICTA assumes that the farmer is a thoughtful individual, who adapts more or less successfully to the difficult conditions under which he farms, and that, in developing technology appropriate to these conditions, professionals have much to learn from the farmer's past experience and ideas. Research and development must therefore be a process in which the farmer participates.

DIGESA continues to operate according to more traditional assumptions. The farmer is still seen as an irrational individual, who cannot be trusted to further his own best interests. He therefore needs direction and close supervision by technicians who propose farming practices that will benefit him. DIGESA's orientation is illustrated by its role in the supervision of BANDESA's credit program. For a farmer to qualify for credit, he must sign a contract to follow a plan worked out for him by a DIGESA agent. The DIGESA agent helps the farmer deal with BANDESA and then supervises his work to make sure that he sticks to the plan.

DIGESA does most of its extension work with farmers on a one-to-one basis. This extension model necessarily limits the number of farmers reached with technical assistance. In administering and supervising BANDESA's loan program, a DIGESA agent can serve only forty-five or fifty farmers a year. With approximately five hundred agents, DIGESA can serve only about twenty-five thousand farm families in one year. To reach even half the number of farm families in Guatemala, this model would require five thousand agents—obviously beyond the nation's financial and managerial capacity. Furthermore, we hear complaints that DIGESA's agents give so much attention to credit that they have little time to attend to the technical needs of the farmers.

Recognizing the need to develop a new extension model to link up effectively with its new research model, ICTA began small-scale research-extension projects of its own. The village of San Martín Jilotepeque, near Chimaltenango, was the first site for this new thrust. World Neighbors, an international self-help organization, had begun a project in San Martín in the early 1970s, following a disastrous earthquake. By the time ICTA came in contact with the project, farmers were already working in organized groups and experimenting with agricultural innovations on their own. They had begun to increase production through the use of soil and water management practices that World Neighbors had introduced.

ICTA and World Neighbors' farmers came to an agreement in 1974 to cooperate in testing some of ICTA's technology in the community. Organizational arrangements for implementing the research were not clear at first, but they have gradually been solidified and formalized. Three informal farm leaders from the World Neighbors group were put on ICTA's payroll, to collaborate with an ICTA technician assigned to the project in managing agronomic trials throughout the community.

This arrangement overcomes the limitation of extending technology to farmers on a one-to-one basis. Professionals can deal with farm leaders, and the farm leaders take responsibility for communicating information and managing experiments with organized groups and communities. Farm leaders in the San Martín project are able to manage approximately sixty field trials a year, compared with the average of twenty-five for the professional *ingeniero* agronomists in the ICTA program. Of course, these trials are not as neat and scientific as ICTA's more controlled experimental work, but they provide data to the research program, and they are of high credibility to the participating farmers. The farmer paraprofessionals working with ICTA extended their work into eleven villages, providing one-on-one technical assistance and holding regular instruction and discussion meetings with the villagers. By 1979 the paraprofessionals were working actively with two large farmer cooperatives, thus further extending their outreach through linking up with indigenous organizations.

Stimulated by the success of the San Martín program, two ICTA *peritos* in the Quetzaltenango region recruited and trained local farmers in the planning and implementation of on-farm experiments. Working with six paraprofessionals in an adult education program financed by the Ministry of Education, one *perito* directed a program of 141 on-farm experiments in a single year. Working with six unpaid leaders of a cooperative, the other *perito* managed a program of 119 such experiments.

This sudden expansion grew beyond the capacity of ICTA to make systematic observations and measurements of yields in all cases. ICTA leaders are enthusiastic over the value of the data acquired, however, and regard the experience as a challenge to ICTA to develop methods of observation and measurement to cope with the expanded volume of experiments, which promise to shape the pattern of field activities.

The cost-effectiveness of these *perito* and paraprofessional projects is impressive. ICTA was paying the San Martín community leaders less than one-half of a *perito's* starting salary and less than a quarter of that paid the *ingeniero*. A skillful combination of professionals, technicians, and paraprofessionals makes it possible to multiply on-farm experiments while adding little to the cost of the program. Further-more, no one who visited San Martín Jilotepeque could fail to be impressed with the enthusiasm and sense of mission displayed by the community leaders working for ICTA. Similarly, the *peritos* in Quezaltenango reported that their paraprofessionals showed great pride and dedication.

In some countries, agricultural ministry planners have been devising systems of hardship pay to lure professionals and technicians into areas far from the conveniences of modern city life. These ICTA paraprofes-sionals require no such inducements. They are where they want to work. They are respected by their fellow villagers, by whom they are selected, and they gain the satisfaction of enhancing that respect as they serve their community (Esman, Colle, Uphoff, and Taylor, 1980).

With such programs ICTA was in effect bypassing DIGESA and justifying its activities as fulfilling its legal mandate to promote the use of new technologies it developed. Whatever the justification, this ap-parent duplication of extension activities raised basic policy questions regarding the responsibilities of the two organizations.

Early in 1978 the directors of ICTA and DIGESA signed a letter of understanding laying out general guidelines for cooperation between the two agencies. By the middle of that year the cooperation in two regions had assumed the form of having ICTA offer classes to DIGESA agents. The curriculum, developed jointly by ICTA and DIGESA coor-dinators, trained extensionists in the practical aspects of conducting experiments with farmers. By 1980 ICTA leaders reported that several ICTA professionals had been appointed as regional directors of DI-GESA. Especially in these regions they found evidence of improve-ment in the relations between the two organizations.

So far such improvements have depended upon the informal initia-tives of individuals on both sides. It now seems clear that no resolution of the problems between research and extension can be achieved short of a basic structural change to bring ICTA and DIGESA under the same leadership. Furthermore, although many DIGESA agents should be able to learn how to carry out on-farm experiments with active par-ticipation of farmers, as long as they carry the responsibility for the supervised credit program they will be able to make only token efforts in the new direction laid out by ICTA.

PNIA in Honduras

The Programa Nacional de Investigación Agropecuaria (PNIA) in Honduras is noteworthy for the following reasons: (1) PNIA is developing a strong interdisciplinary program with major emphasis given to on-farm research with the active participation of small farmers. (2) PNIA is building its program with substantial influence from other national and international programs and with the active collaboration of professionals of foreign agencies but with strong national leadership that is developing a distinctive Honduran model. (3) The research program of PNIA has been influenced by ICTA, but PNIA has gone beyond ICTA in developing effective relations between research and extension. (4) In Honduras research and extension are working closely with effective peasant movements.

Honduras is the poorest country in Central America, yet it may have certain compensating advantages facilitating the development of PNIA. Honduras has a smaller, and therefore politically weaker, elite of large rural landowners, so that the discrepancies in the distribution of income are not so marked as in the other Central American countries (if compared before the Nicaraguan revolution). At the time the new direction for PNIA was established, Honduras had far fewer trained agricultural professionals than other Central American countries, but this lack of enough personnel to establish a strong agricultural bureaucracy seems to have made it easier to strike out in new directions.

Recent developments in PNIA's strategy have been strengthened by a major land reform program, a 1974 decision to decentralize the research organization and build up regional units, and a CIMMYT-Cornell graduate fellowship program, emphasizing interdisciplinary research.

Land Reform and Peasant Organization. The Honduran program of land reform followed shortly after the Punta del Este meeting at which all of the representatives of Latin American governments pledged themselves to land reform at U.S. urging—but only Honduras responded very seriously. Beginning in 1962, the program built up to a substantial rate of distribution by 1964–1965. Fifteen years later, 185,000 hectares had been distributed to approximatey 30,000 families of peasant farmers. Peasant organizations are demanding that the government still continue the land reform distributions, but their claims call for distribution of only an additional 30,000 hectares, about one-sixth of the area already distributed. Clearly, the government has already gone a long distance toward realization of a comprehensive national land distribution program.

The peasant movements had their origin in nationwide strikes against the United Fruit Company and the Standard Fruit Company in 1954. Up to this time, strikes had been illegal, but this conflict of Honduran workers against the foreign banana-producing and marketing companies excited broad popular support and brought about the legalization of strikes. Although the strike leaders sounded radical to their conservative observers, they phrased their demands in terms of indigenous populism rather than foreign communist ideologies, which made it easier to integrate the fruit company workers' unions and the peasant movements into the fabric of Honduran society.

According to a USAID agricultural sector assessment in 1978, the Asociación Nacional de Campesinos Hondureños (ANACH) claimed to have eighty thousand members. The Unión Nacional de Campesinos (UNC) claimed thirty thousand and the Federación de Cooperativas de Honduras (FECORAH) claimed six thousand. An additional estimated eighteen thousand to twenty-two thousand workers on banana and sugar plantations were unionized and naturally sympathetic to the peasant organizations. Some indication of the power of the peasant organizations is illustrated by the political elite's response to peasant expressions of concern that the land reform program was being slowed down. In 1972, the peasant organizations carried out a massive *campesino* march on Tegucigalpa, which precipitated a military takeover of the government. The fact that the military leaders had come to power in response to the peasant movements inclined them to work with peasant organizations.

1974 Regionalization of PNIA. Up to 1974, the agricultural research program had been highly centralized, and many of its critics, including influential people within PNIA, were convinced that it was doing little to help the small farmers. This conviction led to a reorganization of PNIA with emphasis on regional production centers, much as in Guatemala. The 1974 regionalization thus placed PNIA in a position to work more directly with small farmers, but the methodology for carrying out such on-farm research remained to be worked out.

Interdisciplinary Research. An unusual interdisciplinary research program which contributed to PNIA's evolution stemmed from the conviction of the director of CIMMYT's maize program, Ernest Sprague, that conventional graduate education programs in agriculture failed to equip professionals with the ability to work together across disciplinary lines. Since he believed this ability essential to the success of his own program and to good work in international research centers, he persuaded the Rockefeller Foundation to support an experimental doctoral thesis research program for half a dozen students, ranging across a number of different specialties. Sprague undertook to interest several U.S. universities in this program, at first without success. He found a generally uninterested reaction, with the professors refusing to believe it possible to provide really solid education in their specialty if the student was also involved in activities carried on by other specialists. Sprague found this negative reaction initially among several professors at Cornell, but there was sufficient support for the idea to overcome this resistance, and a CIMMYT-Cornell program was set up on an experimental basis.

The six graduate students receiving fellowships ranged across disciplinary lines from entomology, plant pathology, biometry, and agronomy to agricultural economics. Each student spent eighteen months in field work at CIMMYT, concentrating upon research in his particular specialty but meeting regularly with other members of the group to discuss their work. In this way they gained an understanding of the way the various disciplines related to and might complement each other in their thesis work and in working as professionals in a national or international research project in the future. The plan also called for bringing each student's major professor to CIMMYT for joint discussion with CIMMYT staff members and students regarding the group thesis research project on various aspects of maize.[2]

Following the completion of their thesis research in Mexico and before returning to Cornell, the six students traveled through several Central American countries. Their visit to the minister of natural resources in Honduras convinced him that interdisciplinary collaboration was essential to the research program he was trying to develop. On the spot, he offered positions to all the team members. Three of them accepted, including Mario Contreras, who returned to his native Honduras to become director of PNIA after completing a doctorate in plant pathology at Cornell.

Integrated Rural Development with PRODERO. The experience of the Programa de Desarrollo Rural Occidente serves two purposes in this chapter. It is a successful case of integrated rural development, and it provides illustrations of the integration of research and extension. Our interpretation of PRODERO is based upon field work carried out by Lynn Gostyla in July and August 1979.

PRODERO began operation in 1978 in the western region of Honduras, bordering on Guatemala and El Salvador. There seem to be two main reasons why the government chose this region. First, the war with El Salvador in 1969 created a serious concern for the security of that border. Honduran leaders recognized the dangers facing an underpopu-

---

[2]This aspect of the program seems to have had a very constructive effect on the Cornell campus. Some of the professors reported that they had to travel to Mexico to gain an understanding and appreciation of the work of colleagues in other disciplines.

lated and underdeveloped area that was losing population right next to an overpopulated neighboring country. Second, any government dedicated to rural development could not fail to recognize the longstanding neglect of this region, which had far less in government services and physical infrastructure than most other parts of Honduras.

PRODERO was financed in large part by a $20 million grant from the Fondo Internacional de Desarrollo Agrícola (FIDA). Plans called for a five-year program, guided for the first two years by an international Organization of American States (OAS) team collaborating with Honduran counterparts. OAS personnel were to phase out of the project and move on to a similarly conceived integrated rural development program in another region of the country, so that CONSUPLANE, the national planning agency, progressively assumes responsibility for the program.

PRODERO is an unusually broad-based and encompassing program. Agricultural organizations play leading roles, and regional representatives of the ministries of Public Works, Health, and Education are also fully involved. PRODERO has its own headquarters in the same building where leaders of various government agricultural and nonagricultural organizations work and have their offices. Since these Honduran PRODERO officials continue to carry certain responsibilities within their ministries for the region, considerable coordination efforts are required to maintain the integration of the program. The executive force behind the coordination is the Junta de Desarrollo Regional (JDR), which is headed by a military officer, who, under this military government, has the authority to resolve disputes among the agencies involved in PRODERO.

PRODERO aims to work with and through the local social structure. Although PRODERO, in collaboration with the land reform ministry has done work with the "reform sector" controlled by the two peasant organizations, ANACH and UNC, PRODERO has been able to move faster outside this reform sector, working with villages as natural social units. It works with agricultural committees (*comités agrícolas*), where they exist, and stimulates their organization elsewhere.

The main emphasis of PRODERO has been raising agricultural productivity. To support this goal, there is a regional agricultural committee composed of representatives of the various agricultural agencies of the Ministry of Natural Resources.

The OAS person responsible for the research program was Robert Hudgens, a plant scientist, who previously had experience with CIAT in Colombia. Since the only previous research in this region had been done in conjunction with soil surveys carried out in 1977, the program had to be built almost from scratch. Hudgens was at work for six months before it became possible to place a Honduran counterpart with him. He was then able to bring in two Hondurans who had been working with the national maize and bean program, but this background was not all positive. That program had been operating more or less independently of the ministry and had been developing a high input-cost program appropriate for fertile valley land but incompatible with the long-term farming systems strategy being developed by the national research program and preferred by Hudgens for the western part of the country. By August of 1979, Hudgens had seven Hondurans working with him, all *ingenieros agrónomos* who had completed university instruction but had yet to finish their theses.

One of Hudgens's first priorities was to establish an experiment station. He recognized the danger of creating a situation in which research would center on the station at the expense of collaborative and participatory research with farmers on their own fields. But he believed that an experiment station was necessary to carry out certain important research projects requiring a degree of control that could not be achieved on farmers' fields. He also believed that the government would be less likely to abandon research in this region if there was a physical structure, with personnel, to be financed and maintained.

How elaborate should such an experiment station be? The national research program favored a small station with minimum physical investment, but the earlier plans of FIDA favored a larger and better-equipped center. Hudgens's plan, now being implemented, represented a compromise between these two positions.

We will describe the on-farm research process later because that discussion can best be understood following presentation of the new role of extension. In ICTA in Guatemala, the surveys, or *sondeos,* designed to provide agrosocioeconomic information about the area in which agricultural research is to be carried out, are implemented directly by research personnel of ICTA. In Honduras, extension staff have taken over this responsibility.

Although these surveys have clearly been influenced by the ICTA model, Honduran extensionists have taken a different approach to information gathering. ICTA *sondeos* are carried out in a given farming area by teams of agronomists and social scientists. Team members go out into the field in pairs, each pair consisting of an agronomist and a social scientist, and they gather their information through interviews with individual farmers. Team members meet each evening to compare notes on what they have found and to discuss problems and procedures.

In Honduras, for a *sondeo* of six villages in the same valley, for example, the extension service sends teams of four to six members, with each team being responsible for one village. The team meets with groups of villagers—presumably members of the local *comité agrícola*—to discuss the schedule as a group. The Honduran *sondeo* is problem-centered, in that it provides little agricultural production information but rather gives a quick overview of the principal problems seen by the villagers. Furthermore, the Honduran *sondeo* goes further than the ICTA model in gathering information beyond agriculture. The Honduran model is designed to provide demographic information (population, number of farms, number of houses), health information (presence or absence of a local clinic, most common diseases), education (presence or absence of a school building, presence or absence of a teacher, condition of the school building and furniture), local organizations, offices of government agencies, and a listing of agricultural needs and problems. The survey also gathers information on local infrastructure (presence or absence of access road, electricity, potable water, and latrines).

The field team is able to fill out this schedule inventory in a couple of hours of discussion with the local farmers. Each team prepares a written report. Then the regional director calls the teams together to meet with key research and extension staff and communication media officials. In the area of La Empalizada, for which we have the summary report of the field teams, the oral discussion of each village report was led by team members, in a meeting attended by thirteen officials in agricultural research and extension.

From such a meeting extension and research planners learned, for example, that only one of the six villages had a problem of lack of sufficient land for maize, whereas all six villages had serious problems with slugs in growing beans. In other words, the summary report tells the planner what problems are general throughout the area. This information helps extension and research to determine priorities in their programs. In this case, if a treatment of the slug problem is known and available in Honduras, extension can seek to provide the farmers of this valley with the needed information and try to help them to acquire the necessary materials. If a method for effective treatment is not known, at least for this particular area, research planners are prompted to give serious consideration to this problem.

Before research activities were begun in the western region, extension people had been frustrated, finding that their traditional approaches were not effective. They came to recognize that they were working in a region quite different from areas from which the standard recommendations for raising corn, beans, and other crops had come.

The western region is very hilly, with many farms located on land with a 30 percent grade. These lands suffered severly from erosion, so the application of fertilizer and other modern inputs generally was a waste of money because the rains would wash them away.

Two of the extension agents in the western region had previously attended seminars conducted by Marcos Orozco, a Guatemalan specialist in soil conservation. The Honduran agents now decided that in this particular region, soil conservation must be the foundation for any agricultural production program. They sought out a farmer who was working land on a 30 percent slope next to a main road, which would make any change highly visible. They persuaded him to work with them in applying the Orozco conservation methods to part of his land. The results were spectacular. The experimental plot yielded ten times the production of the control plot. The farmer and his friends were impressed, and the word began to spread.

The extension agents went on to organize small independent farmers into committees for soil and water conservation. They invited Orozco to teach farmers his simple but effective methods. They also arranged to have groups of farmers travel to Guatemala to visit the World Neighbors project being carried out under the direction of ICTA by paraprofessionals in San Martín de Jilotepeque. The farmers returned from San Martín with enthusiasm, and the word continued to spread.

The next step involved a distribution of fertilizer and improved seeds. Following Hurricane Fifi, Honduras received substantial international aid for agriculture from 1974 into 1976. Some of this material was still available in early 1978. In the past, the extension procedure had been to give away the inputs, thus providing the favored farmers with a benefit for one year but leaving them dependent upon credit and the availability of supplies if they wanted to use the same level of inputs the following year. In the western region, apparently stimulated by the leaders of PRODERO, extension carried out a different plan.

The cooperating soil conservation committees were renamed *comités agrícolas* to indicate their broader scope. Inputs were not distributed to individual farmers but were given out in each village to its own agricultural committee. Extension also announced that new committees could be organized in order to get these inputs.

Extension laid down three conditions for receiving this benefit: (1) All members of each committee must enroll in a one-week course on conservation methods and agricultural practices. (2) The inputs were to be used only on land where the methods being learned were to be applied. (3) At harvest time, members would pay back to their own group an amount equal to or greater than the value of the inputs they had received. They would then be in a position to purchase the necessary inputs for the next season. (The money was deposited in the agricultural bank for the period between harvesting and planting.)

The results were impressive. The ten extensionists operating at the start of the program could not keep up with the demand. (Their number was to be doubled the following year.) Demand was so high that the inputs ran out. Robert Hudgens was able to get $3,000 from CATIE to purchase more inputs. When that additional supply ran out, Hudgens and the extension officials prevailed upon AID to put up $15,000. The money came too late for that planting season but was used for the next one.

Extension had planned to work with or organize twenty groups the first year and ended up with thirty-eight, each group having fifteen to forty members. By the second year, extension expected to work with fifty-two groups, but, handicapped by the late arrival of money for inputs, ended up with forty-two.

Extension is now planning to extend its own reach through working with paraprofessionals, known as *auxiliares*, local farmers chosen by the groups themselves, each one working with several farmer groups. According to the new plans, the extension agents will meet with their village groups, one day for each, Monday through Thursday, and have training sessions with the *auxiliares* on Fridays. The *auxiliares* are not to be paid a salary but are to be compensated for time lost from their own farm work. Some *auxiliares* are already functioning informally, immediately after participating in the group discussion and training sessions. Extension is now developing the same methodology on the *asentamientos* of the reformed sector, grouping several *asentamientos* together and having them organize their own agricultural committees.

PRODERO has also worked out new arrangements for farm credit. Previously small farmers in this region could not get credit on a group basis without having legal recognition of their group, without personal identification papers for all members (many have no birth certificate or similar paper), and without using their land as collateral. PRODERO has persuaded the agricultural bank to abandon all of these requirements and simply use the prospective crop as collateral. The groups are now banking their money. The bank loans its own funds at 6 percent interest to groups, which in turn lend to their members at 11 percent. If the plan succeeds, this 5 percent difference will become part of a rotating fund, which can be used to install a supply store, a dryer, milling equipment, and so on. The idea is to have the group own such facilities and rent them at a low rate to members and at a higher rate to nonmembers.

The Research-Extension Process. Since no agricultural experiment station is yet available, plant research in the western region necessarily begins with farm trials. The first trials were carried out on fields of small farmers, the principal target population for PRODERO. Hudgens concluded, however, that it was not feasible to carry out the variety of treatments planned on very small farms. The process now moves through the following steps:

(1) Farm trials on fields of the larger farmers.

(2) Preparation of packages for maize and beans at three levels of input technology.

(3) Giving the package to extension agents for work with the agricultural committees.

(4) Planting in small plots by members of the agricultural committees, with the guidance of extension agents. Since small amounts of material are involved, with a number of farmers working together, this part of the process takes only about a half hour of work in the field.

(5) Weekly meetings of each agricultural committee with its extension agent to check on progress and discuss problems.

(6) Research people come, especially toward the end of the growing season and at harvest time, to check on results and to discuss them with farmers and the extension agents.

The experiments carried out so far with maize and beans have involved comparisons between the native or *criollo* varieties with six treatments for each at three levels of inputs.

Farmer Participation. Although our field work has not been extensive enough to measure the extent of small farmer participation in the Honduran program, Gostyla's interviews and observations suggest a very active level of participation. Even in activities under the direction of a professional, the aim is to achieve a balance in the presentation of information by the professional and the farmers reporting and discussing their experiences. For example, extension conducts a radio program every morning from 5:45 to 6:15. A part of this program is devoted to interviews with farmers. Going into the field with a man in extension communication, Gostyla was surprised to find herself participating in the next morning's radio program—the extension man recorded her interview with a farmer.

Twenty-four farmers were present at the meeting Gostyla attended of an extension agent with an agricultural committee. Since the members represented different levels of education and length of experience with the agricultural committee, it was not an easy task to achieve full and free communication, but the extension agent was able to stimulate such participation. The focus of this meeting was on presentation and discussion by group members of successful and unsuccessful techniques used during the past agricultural cycle.

The agent took pains to move the participants beyond general statements that might simply reflect what they took to be the accepted extension doctrine. He emphasized that he wanted them to report their real experiences. At the close of the meeting, the agent promised to write up what they had told him and come back for the next meeting with a draft of the report. After they had discussed it, he would make revisions and come back again with a mimeographed report, so that each member could have a copy of what had been learned.

Gostyla found many farmers learning from other farmers and from professionals in other countries. One older farmer, a member of the committee whose meeting Gostyla observed, reported that he had been to San Martín Jilotepeque and had returned very enthusiastic. He had been particularly impressed by the demonstration that there were real and attractive alternatives to traditional practices. Gostyla also encountered a sixteen-year-old farmer who claimed he was getting impressive results with what he had learned locally and in Guatemala. Again the San Martín visit seemed to have an important consciousness-raising function. The young farmer reported that he had learned much from his father and was now able to teach his father some things. Gostyla was also impressed with the farmers' understanding of the logic of the experiments in which they had participated. She interviewed one member of an agricultural committee that had started with eighteen members and now has thirty-eight. At the time, he was filling out papers to go on a trip arranged by extension to the Instituto de Agricultura Indígena in Mexico.

This farmer said that he had earlier used fertilizer but had stopped when he felt that it was not working. Now he reported that combining fertilizer with the conservation methods he had learned had more than tripled his maize yield. The farmer was able to explain in detail the six treatments of maize and of beans in the experiments on his land, stating not only the procedures followed but also the logic underlying each step. He added that so far he was getting better yields with his *criollo* maize with all inputs than he had achieved with the improved maize under the same conditions, but he was planning to test further to make sure he came out with the best combinations.

Problems of Interorganizational Relations. The success of the Honduran program depends in large measure upon development of effective relations between international and local agencies and also upon the coordination of agencies in Honduras. In the early months, PRODERO experienced some difficulties in gaining the collaboration of personnel from various ministries. CONSUPLANE, the national planning agency, had overall responsibility for guidance of this development program, but CONSUPLANE is not an operating agency with an active presence in the region. When interagency disputes arose, it was sometimes necessary to appeal to the military governor of the region or to Tegucigalpa for decisions. As the project progressed, such outside appeals for collaboration decreased.

The withdrawal of the international professionals from PRODERO after late 1979 and the ending of military rule naturally raise the question of whether the preexisting level of interagency collaboration can be maintained, or whether each agency will simply go its own way. Since the Ministry of Natural Resources is now more strongly represented in both research and extension in the western region, the government has decided that it will assume the leadership and coordination responsibility.

Linking Farmer Organizations with Government Agencies. In recent years, Honduras has moved rapidly in a program of linking farmer organizations with government agencies having responsibilities in agricultural development. The initiative on the government side was in the hands of Rolando Vellani of FAO, working with a small group of Hondurans. They started in 1975 visiting forty to forty-five production cooperatives that were based upon peasant organizations, seeking to determine the factors associated with their success or failure. As they interviewed farmers and attended cooperative meetings, they came to recognize that, although ecological conditions were of some importance in the effectiveness of the organizations, they did not seem to be the determining factor. Some cooperatives enjoying very good ecological conditions seemed to be failing, while others struggling under poor conditions were succeeding. Finding that the skill of organizational leaders in the internal administration of the cooperative and in its external relations was of great importance, they realized that they should give special attention to ways of developing effective relations with government agencies, a feature that was markedly lacking throughout Honduras. If they were able to link cooperatives more effectively with government agencies, the way could be opened to technical assistance in which government officials could help cooperative leaders to improve their internal administration.

Over a period of two years, Vellani and his group worked closely with seven cooperatives in the area of Los Almendros. This experience further convinced them of the importance of external institutional factors. They found, for example, that farmers had great difficulty in getting credit in time to make optimal use of the inputs they purchased.

During this period, leaders of the seven cooperatives began to meet monthly at Los Almendros to discuss credit, production, and internal management problems. In February 1979, these cooperative leaders invited Vellani and his associates to help them plan production for a newly formed regional agricultural and cattle cooperative, made up of eighteen to twenty local organizations. Vellani accepted this invitation for his own group and also arranged for regular Monday meetings of the regional cooperative leaders with the regional heads of agriculture-related agencies: the ministry of natural resources, the agricultural bank, the land reform agency, and the government agency charged with assisting cooperative development. The production plan for the first year covered 700 hectares to be planted to maize and beans (beans taking up about 210 hectares). They did not at this stage work out an intercropping program but planted each crop separately.

On the basis of information locally available from agricultural researchers in the Ministry of Natural Resources, Vellani and the regional leaders worked out a production plan to present to the bank. They figured the costs of improved seeds, herbicides, fertilizer, and fungicides. They also estimated what they would need from the bank to buy tractors and their attached implements, to be owned by the regional cooperative. The cooperative now has its own pool of machines, hires its own mechanics, and has trained tractor drivers. Each individual cooperative contracts with the regional cooperative for tractor and driver.

The first regional cooperative arranged to finance its crops with a $250,000 bank loan, an amount ten times what any local cooperative previously had been able to secure—and this time the loan money was available when needed. It also arranged a $250,000 bank loan to purchase equipment. The regional cooperative then constructed a building for the machines and also for crop storage.

It is now becoming a policy that 1 or 2 percent of the amount borrowed by the regional cooperative for inputs and equipment will be added to the loan to cover costs of administration. In the regional cooperative, administrative costs are kept at a minimum, with most services performed by unpaid officers, but there is of course a need to compensate the regional leaders for travel and living expenses for their cooperative meetings and for their consultations with government officials.

When we visited Honduras in early 1980, the Vellani group had not yet finished analysis of the results of the previous crop year, but preliminary indications pointed to a yield of maize more than double that of the previous year. Vellani reported that weather conditions and especially the amount of rainfall had been practically identical for the two years.

The success of this first regional cooperative effort attracted widespread attention as news spread rapidly through the ANACH organization. By early 1980, six other regional cooperatives had been formed, and Vellani noted growing interest throughout the countryside in moving beyond the local cooperative base toward regional organizations.

February 1980 marked another major step organizationally as ANACH and the government began regular monthly meetings in Tegucigalpa between the operating heads of all agriculture-related agencies and the leaders of the regional cooperatives from various parts of the country. If we consider the formation of a local cooperative as a first-level organization and the organization of a regional cooperative as a second-level organization, this program of monthly meetings constitutes a third level of organization. The farm leader involvement at this third level provides the organized small farmers with channels of influence on broad questions of agricultural policy and on administrative problems of government agencies serving the rural population.

Vellani emphasizes the importance of the production plan as a basis for economic advance and organizational development. Working out the production plan gives the regional cooperative leaders and the Vellani group valuable experience in gathering and interpreting information and developing the organizational base to put that information to use. The successful execution of the production plan brings financial rewards that strengthen the commitment of members to their local and regional cooperatives and give the regional leaders the sense of success that encourages their further work.

The organizational development described here seems of particular significance in building linkages between the small farmers and government agencies and in securing accountability of those agencies to the base organizations they are expected to serve. This three-level organization of cooperatives, linked with government officials at every level, helps to secure coordination of action of the bureaucracy with the needs and interests of the peasants. The organization also provides pressures toward government accountability, as local and regional leaders push for service improvement in all fields.

Vellani has observed a rapidly growing demand for training being voiced by regional cooperative leaders. Contrary to the situation that traditionally prevails, in which government bureaucrats themselves decide what training peasants need and then go out and try to persuade peasants to submit to such training, in Honduras regional cooperative leaders articulate the needs that they and their members feel. They thus have some assurance that the training the government offers will meet really felt needs. Therefore, it is likely to be more valuable to the farmers than any training that government officials simply tried to sell them.

For example, the farmer leaders have made it clear that they and their members have been persuaded of the advantages of cooperation and do not need further emphasis upon its philosophy and ideology. They are more interested in training in accounting and recordkeeping, production planning, and other technical subjects.

This three-level organization has enabled the small farmers to achieve enormous economies of scale at minimal cost to the government. The cooperatives manage themselves and operate with their own funds. The cost of administering the regional cooperatives and of financing the meetings of their leaders in Tegucigalpa is covered by bank loans, which the regional cooperatives repay.

Economies of scale are also offered to the government through its dealings with the organized farmers. To be sure, the effectiveness of the farmer organizations may require government to make expenditures and undertake activities beyond those planned in a traditional organization. But these additional expenditures are more than counterbalanced by savings achieved as research, extension, and the agricultural bank no longer have to deal with individuals or small groups. In regions where cooperative organizations are strong, government officials provide information, financing, and technical assistance to the regional leaders, who manage the physical facilities and human activities required for efficient farm production and marketing.

If the government were required to hire and train its own employees for all of the functions performed by local and regional cooperative leaders, the costs would be enormous, and the results would be far less satisfactory to the people being served. Also, the experience local and regional leaders are gaining enormously enriches the human resources developing in the countryside, as the peasants organize to meet their own needs instead of depending on paternalistic government.

Conclusions

Having traced in earlier chapters the evolution of some of the basic ideas underlying participatory strategies for agricultural research and development, in this chapter we have examined the implementation of such strategies in two national programs. As noted, implementation requires solving two basic problems of organization structure and social process: (1) to devise a system of on-farm research built upon the active participation of small farmers, and (2) to integrate the on-farm research program into the already established national programs of experiment stations, extension, credit, and marketing.

In Guatemala, ICTA developed a promising new system of on-farm participatory research with small farmers. As we have seen, this new approach was not invented all at once and then applied in the field. ICTA people began with a general sense of direction and then designed and redesigned the system to include active involvement with small farmers.

The ICTA system has innovated in adopting a nontraditional starting point for on-farm research. Instead of starting by trying out on farmers' fields the ideas and technologies developed on experiment stations, ICTA staff began with intensive field studies of the farming system currently being practiced by small farmers.

The utilization of such research by staff outside of the socioeconomic unit was initially held back by two factors. First, the studies were initially carried out by organizationally separate social scientists, which naturally reduced their credibility with plant scientists, who dominated ICTA. Second, the socioeconomic unit took a full year to get from the beginning of its farming system studies of an area to the submission of its final report. Therefore, plant scientists could argue that, even if they accepted the scientific soundness of the unit's studies, the methods used were too slow and not cost-effective.

ICTA resolved both problems by forming integrated field survey teams of social and natural scientists and by developing an abbreviated methodology that provided area data from a two-week field work period. Leaders of ICTA concluded that the shortcut method produced adequate data for project and program planning.

We have traced the sometimes difficult and awkward process of fitting this new element of on-farm participatory and interdisciplinary research into the previously established national programs. To achieve its full potential, this new style of research had to progress through a series of organizational changes, with the socioeconomic unit initially attached only to national headquarters and then fitting itself into the regional level of organization.

In Guatemala, one major problem remains unresolved: the disfunctional relationship between an innovative research organization and a traditional extension organization. Such ineffective relations are all too common elsewhere in developing nations. This problem was addressed differently, and resolved, in the Honduras case, where the parties have made great progress in improving the research-extension relationship.

The leaders of PNIA profited greatly from their study of the ICTA model, but they did not simply copy it. Well aware of the deficiencies in the research-extension relations in Guatemala, PNIA leaders developed field operations in which extension agents were no longer passive recipients of ideas and information furnished to them by researchers but became active participants—together with farmers—in the farming system surveys. In examining PRODERO, we have seen Honduras carrying out an integrated rural development program, within which agricultural research and extension play leading roles but work closely with regional officials and active villagers on problems of credit, marketing, health, and education.

The two cases also demonstrate the importance of local-level farmer organization in agricultural R&D. Regardless of the quality of the interpersonal relations involved, a one-on-one relationship between the agricultural professional and the small farmer is inherently inefficient. It is far more cost-effective for the agricultural professional to work with and through an organized body of small farmers. ICTA took important steps in this direction but was limited by the jurisdictional claims of DIGESA for responsibility for diffusion of innovations. Leaders in Honduras profited both by a much more effective research-extension relationship and by local organization based upon widespread and active peasant movements. Instead of regarding such peasant organizations as threats to government—a view common in other countries—Honduran leaders chose to work with and through these grass-roots organizations. Furthermore, Honduras has taken important steps to link these peasant organizations with government agriculture-related agencies at regional and national levels. Instead of limiting agricultural planning to national officials, Honduran officials are beginning to work on collaborative planning with officers of regional cooperatives.

In emphasizing the importance of active farmer participation in the R&D process, we do not mean to exaggerate the capacities of the small farmer or to minimize the contributions of the professional. The farmer is not likely to know what new genetic materials or other inputs may be available or might be provided through further research. But he is the expert on his own farming system and the conditions affecting it and the needs and interests of his family. The professional can bring new materials and new ideas to the attention of the farmer, but they will be useful contributions only insofar as he learns from the farmer how to fit them into the agrosocioeconomic system of the farm and the community of farmers. The participatory approach requires the professional to abandon efforts to "sell" new ideas to small farmers. The professional can serve the small farmer best if they can work together to devise and define new options among which the farmer does the choosing.

While acknowledging that ICTA and PNIA are doing good things, those schooled in conventional styles of agricultural research may question whether the on-farm interdisciplinary and participative projects represent "real research," thus raising a philosophy of science question. What we describe is agrosocioeconomic research—a new hybrid but no less genuine as research. When staff people go out to get data through observing on farmers' fields, interviewing farmers, measuring inputs and yields, and so on, they are certainly carrying out research operations. Furthermore, when extension agents participate actively in this data-gathering process, they, too, are involved in research, whatever their titles and job classifications. In fact, we would argue that the attempt to draw a sharp line between research and extension has negative effects upon both research and extension.

Of course, we are not arguing that this emerging style of on-farm research should be substituted for scientific work in laboratories and on experiment stations. We see the two fields of activity as complementary rather than competitive. Field researchers can offer farmers new options for increasing yields and incomes on the basis of materials and ideas generated on experiment stations. Ideally, the influences will go in both directions as researchers discover promising new farming systems in the field and bring back ideas that require refinement and testing under the tightly controlled conditions of a good experiment station.

Part IV

Organizational Implications

Part III focused on the farm family and the small farm enterprise in the context of the rural community. We were especially concerned with the interests, needs, and motivations of small farmers and with the discovery of participatory R&D systems that are more responsive to those interests, needs, and motivations.

In Part IV we are focusing upon some of the dilemmas of size and structure facing small farmers and those seeking to help them. The small scale of their operations provides these farmers with both advantages and disadvantages. From this standpoint, the problem of development is to find ways of retaining the advantages and minimizing the disadvantages.

On the favorable side, various studies have shown that small farmers in general make somewhat more efficient use of their (limited) resources than do large operators. They learn to know the characteristics, the potentials and limitations, of every square meter of their land. Depending little if at all upon hired labor, they discipline themselves to provide for their farms the "tender, loving care" we hope to find in the nurse-patient relationship. Furthermore, since they themselves make the farming decisions, they can adjust rapidly to changing conditions and new opportunities.

On the unfavorable side, individual small farmers are unable to take advantage of the economies of scale open to large operators. They are at a disadvantage in the market place both in buying their supplies and equipment and in selling their produce. They often lack the money to buy what they need to increase the productivity of their farms, and they have problems in getting credit promptly and on favorable terms. They may recognize long-term needs for soil conservation and yet lack the resources required for such measures. Their social position and limited resources place them at a disadvantage compared to large operators, when they seek help from politicians or administrative officials.

Individual small farmers can only achieve economies of scale if they build an organization responsive to their needs and interests. We explore this aspect of development in Chapter 14, "Farmer Organization and Participation as Keys to Agricultural Research and Development."

Building an effective farmer organization requires not only developing an internal system of interpersonal relations and linking it with the external social environment. It also depends upon linking the social system of farming with the needed technology and work requirements—in other words, it depends upon building an integrated socio-technical system. Chapter 15, "New Directions in Irrigation Research and Development" examines a socio-technical system built upon the technology and work requirements of irrigation systems.

Government bureaucracies exhibit their own dilemmas of size and structure. If they are designed to meet the needs of farmers throughout a nation, they are necessarily large organizations. In the past, these large organizations have suffered from rigidity, poor coordination among various agencies designed to serve the same public, and unresponsiveness to the needs and interests of small farmers. Chapter 16, "Bureaucratic Reorientation for Agricultural Research and Development," describes the characteristic problems of conventional government organizations and goes on to present a strategy designed to improve their performance especially in relation to small farmers.

14

Farmer Organization and Participation as Keys to Agricultural Research and Development

If we are to understand the behavior of farmers, we must view the community and the world as they do. We find that small farmers tend to be concerned not simply with individual material gains or losses but also with the economy of family and household. To understand the social economy of the household, we need to know, among other things, the number of people potentially available for agricultural and other work and their ages and sexes, which tend to determine what kinds of work are possible and appropriate. We also need to recognize that, for many small farmers, income is not limited to what they produce on the farm. Many work part-time on the family land, while other members of the family may have off-farm sources of income.

Within the limits of strength and endurance, farmers tend not to give great weight to the cost of labor which family members put in on their own farms—except when the need for additional labor would mean giving up other earning opportunities or would require hiring non-family members. If a change in the management of the farming system would require substantially more labor than has been customary in order to achieve increased yields and income from the farming activity, we need to consider whether enough family labor would be available to do the additional work. We need also to consider whether, to provide this extra labor, family members would have to give up earning opportunities off the farm. Of course, much depends upon the timing of the additional labor requirement, but farm families cannot rearrange work schedules at will.

It has been commonly assumed that in many developing countries a large labor pool exists that is not fully used on the farms, so that farm families can readily increase their incomes through taking on additional tasks that would increase production. To be sure, there often are slack periods when family members have little necessary farm work, but the potential for introducing changes should not be judged according to the availability of labor in slack periods. Even if the family has more than enough labor available for much of the year, its members may be fully

197

occupied and must even bring in additional labor to work during peak periods such as planting and harvesting. Therefore, in considering the feasibility of any innovation requiring additional labor or changing the timing of labor demands we need to consider not only the labor generally available on the farm but also its use at periods of peak labor demands.

Agricultural economists have usually been inclined to assess the feasibility of employing additional labor for an innovation by looking at the increase in yields and income that might result. They are then often puzzled when farmers fail to adopt an innovation that would seem to offer a marked increase in their farm income. But it may be rational economically for the farmer to reject a particular innovation if it means giving up comparable or better earning opportunities in other activities, if it means extra costs in time and money to hire and manage nonfamily labor, or if it involves serious sacrifices of rest or leisure.

We need to address issues of availability of labor and potential labor shortages because these questions have often been overlooked. The question of capital shortage is much more commonly recognized and therefore needs little discussion. As we have seen, it involves not only the availability of capital, with the small farmer generally at a distinct disadvantage, but also the timing of credit. One so often hears farmers complain that the money they need is available only a month or so after they can make best use of it that one wonders whether agricultural credit agencies have a system that guarantees the late arrival of loan money.

We must also consider the factor of risk when studying farmer behavior in relation to credit. Most advances in agricultural technology require expenditures beyond the farmer's customary level, and some require very substantial amounts of money. In evaluating the cost relative to benefits of new agricultural technology, it is not enough to demonstrate that the farmer will be substantially better off financially on the average. As the importance of risk has come to be more fully recognized, agricultural economists now attach importance to the standard deviation of income from year to year as a way of assessing risk, in addition to calculating average yields and incomes.

When a small farmer is operating without credit, he may grow enough in a good year to feed his family and have a small surplus to sell. In a bad year, he may be hard-pressed to feed his family and have no surplus; still he can adjust to that adversity as his father and grandfather did by reducing consumption or getting consumption loans. If he borrows money to improve production, he may come out substantially ahead in a good year, but a bad year will leave him with little or nothing to market and therefore unable to repay the loan. If he has pledged his land as collateral, he runs the risk of forcing himself (and future generations) into the class of landless laborers.

Recognizing the importance of the family and household as the smallest organized unit of rural society, we need to consider how that unit can be linked more effectively with the socioeconomic and political structures of the area, region, and nation. Our concern here has been to discover how small farmers can be better integrated into the social process of agricultural research and development. We have argued that top-down, paternalistic programs are not likely to be successful in assisting small farmers, no matter how benevolent the intentions of the planners and administrators. Farming systems are so complex and the constraints which farm families need to work within are so numerous that relevant research findings and effective extension advice are unlikely to be forthcoming unless intimately related to the experience, perceptions, and realities of small farmers.

We therefore need to reconceptualize the organizational approach to agricultural research and extension. Figure 14.1A illustrates the structure and social processes of agricultural R&D under a conventional system of developing countries and contrasts it with an emerging new system. Figure 14.1B is based upon the ICTA model in Guatemala, but the principles involved are the same in all emerging systems involving on-farm research with active participation by the small farmer.

Figure 14.1 also indicates how government programs can be restructured so as to facilitate peasant participation. Although an important advance over traditional bureaucratic systems, this restructuring of the research and extension activities of government is not sufficient to provide small farmers with the help they need. Such restructuring of research activities, based upon the voluntary decisions of high officials to allow peasants to get into the act of agricultural research and development, may not have profound effects by itself. The officials who initially developed such a system of participatory research might later be supplanted by others whose interests are linked with those of large farmers and agribusiness or who still see small farmer development problems in traditional trickle-down terms. In such a case, the agricultural and social scientists in the field would lack the support needed to sustain the participatory program.

Whatever the rhetoric of its spokesmen, in a traditional bureaucratic organization the official in the field is accountable to his superiors; how well or poorly he meets the needs of the farmers he is supposed to serve may have little bearing upon his career. Those who please superiors and do not run afoul of powerful interests have the most successful

Figure 14.1. Organizational approaches to agricultural research and development

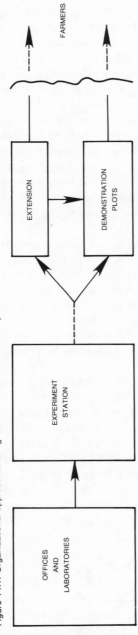

A. Agricultural research and development model based on organizational approaches in developed, industrialized countries

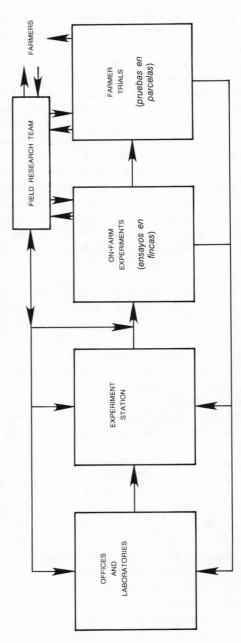

B. Evolving agricultural research and development model based on the approach of ICTA in a developing, predominantly agricultural country

careers. If farmer participation is to fulfill its potential for contributing to social development and economic growth, it must be supported by a major shift in accountability so that the staff in the field are accountable for their performance to farmers as well as to their superiors.

That shift in accountability can be accomplished most significantly through organization and increasing the resource base used by peasants. If the small farmers gain more resources, they become less dependent upon others, including government, and gain more influence on government agencies. Land reform provides one major avenue for increasing the peasant resource base and has been important in doing so in Honduras. Land reform is such a large subject in itself, however, that we cannot deal with it at any length here. Land reform may be a necessary condition for the improvement of the lot of the rural poor in many developing countries, but it is rarely a sufficient condition. Without access to land, there is of course no way that one can be a farmer. Still, making a living for oneself and family largely or entirely on the land requires other resources besides land. If the rural poor are to benefit from land reform, the change in land titles must be accompanied by an improvement in access to the other resources needed to make the land fruitful.

We should recognize that organization is also an element of rural infrastructure, as important as roads or markets, potentially offering economies of scale and mobilization of resources. An effective organization can provide its members with access to credit more efficiently and on a more timely basis than would be possible for the individual members. By pooling members' purchases and sales, the organization can buy more cheaply and market more efficiently and profitably. Organization also makes possible investments in machines, equipment, and buildings that would be beyond the reach of individual members or small groups.

On the other hand, as small farmers organize, they face more complex problems, and many cooperatives have failed because of inability to manage this increasing complexity. The management of cooperatives requires an increase in the quality of the human resources charged with cooperative administration—in other words, members, and particularly organizational leaders, need to learn new skills, and the organization needs to find ways of training its members.

We have traced the evolution of ICTA from its on-farm experimentation with small farmers to paraprofessionals working on research and development with cooperative organizations. With PRODERO in Honduras, we have described the growth of village associations based upon the management of an agricultural credit program. Also in Honduras, we have noted the development of the peasant movement known as ANACH and shown how its local base organizations are being linked in regional cooperatives and how the officers of those cooperatives have established a third-level organization, effectively linking them with government officials at both the regional and national levels. Small farmers are collectively gaining access to power and resources out of their reach as individuals.

In Honduras, we have seen how government officials are able to facilitate the organization of small farmers and aid in linking them with government agencies. The national organization of the peasant movement already existed in Honduras, however, independent of government support and assistance.

We recognize that peasant organizations may be seen by many politicians and government administrators as a two-edged sword. The leaders of nonparticipatory governments may recognize the economies of scale made possible by peasant organization but will be apprehensive about the political potential of such movements. They recognize that a peasant organization formed primarily to pursue economic goals will make demands upon government and may become a potent political force. Even if government leaders are prepared to accept the political potential of peasant organizations, they may not find it easy to intervene in such a way as to stimulate and support them. One of the key problems involves learning how to assist and facilitate without creating dependency. Let us consider here several cases that suggest at least partial answers.

The Fomentadora Rural S.A. (FORUSA) model of rural development corporations provides an interesting test of the possibility of stimulating agricultural development through the private sector. This model grew out of experimental work by the U.S. agriculturist Simon Williams in Mexico and particularly in the Guadalajara region. Having shown promising results from his small-scale experimental work with small farmers, Williams persuaded the leaders of the ICA group, a large Mexican engineering, manufacturing, and construction corporation, to finance a program designed to discover whether it would be possible for a rural development corporation to work with small farmers in *ejidos*[1] and improve their yields and incomes so much that the farmers would ultimately be willing to pay enough for the technical assistance to cover its costs. Those establishing the FORUSA model

[1] An *ejido* is a rural settlement in which families control and farm roughly equal parcels of land. The family can pass the farm down from generation to generation, but it cannot (legally) be sold to outsiders.

recognized that it would not be possible to cover any more than a small fraction of the annual cost of technical assistance in the first years of a given project but hoped that increasing success would eventually generate income to cover its costs.

FORUSA is involved in projects in three communities in the rural area around Guadalajara. In the project most successful there so far, five years after it started, 89 families (out of about 250 families of the *ejido*) had a functioning cooperative served by one full-time agronomist. The three projects jointly were served by an agricultural engineer, responsible for development and improvement of irrigation, and by an accountant, responsible for providing instruction and technical assistance in recordkeeping and financial management, plus a program manager. The cooperative we visited, Zacotipan, had been able to purchase and manage eight tractors, two threshing machines, two large trucks, and two pickup trucks. The purchases were financed by loans from a private bank, arranged with the assistance of FORUSA, with the final payment to be made in 1980.

The farmers in this community, with the guidance of FORUSA, are concentrating particularly on raising sorghum. Each family farms eight hectares and pays approximately $10 per year per hectare for the technical assistance it receives through FORUSA and its cooperative society. Before the start of the agricultural cycle each year, the farmer works out a production plan, with the advice of the FORUSA agronomist. The plan provides in detail for the amount of seed, fertilizer, insecticides, and other inputs to be purchased and also for the number of days of machine work to be contracted for with the society. The total of these figures, plus the technical assistance charge, with the subtraction of whatever the farmer is able to pay, yields the amount the farmer will borrow from the bank that year.

Throughout the growing season, the agronomist is in the village every day observing and consulting with the farmers. When he is called upon to give advice on a particular problem, he not only presents the advice orally but also writes it out and gives the farmer a copy. These technical assistance records, together with the annual record of expenditures, yield, and income from the farm, provide the basis for systematic farm production records. With leaders of the cooperative society, the agronomist visits the market and makes arrangements for delivery and sale of the produce. Each farmer accompanies the truck with his produce to the market and observes the process of weighing his crop and calculating the yield and the amount of money he receives.

When we visited in early 1980, the cooperative was taking the next steps to intensify its economic activities. Members were establishing a feedlot to fatten cattle with their large production of sorghum. They reported that they had more than doubled the sorghum yield and now found it more profitable to invest a portion of this yield in animals rather than sell the entire crop. FORUSA had also worked out with the cooperative leaders an arrangement whereby a manufacturer of trousers would set up a small plant in the village and provide machines and instruction in their operation, as well as materials. The workers in the plant were to be mainly women, wives and children of the co-op farmers. The cooperative was also about to establish a bee and honey project.

At the time of our visit, when the farmers' income was limited to the crops they grew, the technical assistance payments farmers made to FORUSA appeared to cover less than the salary of the full-time agronomist allocated to that village. Thus FORUSA was subsidizing them to the extent of one-third of the time of the three additional officials, the accountant, the agricultural engineer, and the Guadalajara FORUSA manger, plus their share of additional expenses for transportation of staff members, secretarial expense, and office operation in Guadalajara. FORUSA officials expected that the additional economic activities stimulated and guided by FORUSA would steadily narrow the gap between expenses and income, though clearly by mid-1980 they were still far from their goal in the Guadalajara area.

FORUSA officials report that within four years of the establishment of projects in the area of Tampico in the tropical lowlands, they have been able to cover 50 percent of their expenses from technical assistance fees. They give several reasons for the more rapid progress in this area. In the Guadalajara region, they were dealing with established communities, where the poorer villagers have long been dependent upon their more affluent neighbors for loans and other favors. A situation has been created in which the freeing of the villagers from dependence represented a threat to traditional leaders and generated considerable opposition. By contrast, the projects in the Tampico area served a new colonization area whose families, being newcomers, were not yet tied in with economic relations of domination and dependency within their own community. Probably still more important, in the Tampico area each family has twenty hectares. Since the Tampico farmers can fit in two crop cycles a year, whereas around Guadalajara only one is possible, the farmers in the lowlands were in effect farming forty hectares, compared to eight for the Guadalajara farmers.

When we consider the number of hectares involved in the Guadalajara region, we recognize the limitations of the FORUSA strategy. In fact, FORUSA spokesmen do not argue that their strategy will work with the poorest farmers. The minimum size farm that would support a FORUSA model, under varying ecological conditions, remains to be determined. One should not rule out the model for smaller farmers but rather recognize that in some situations the rural development corporation must have a subsidy, which can only be expected to come from the government. In fact, in 1980, the land-tenure and colonization agency in Costa Rica was planning an experimental project whereby the government agency would set up its own rural development corporation, which would provide technical assistance to communities of farmers in an area where a large dam and irrigation project was to increase substantially the agricultural potential of a wide area.

Another example is a government-sponsored project at Zacapoaxtla in Mexico, which grew out of the original Puebla project. The officials there had started agronomic experiments but discovered that the farmers had so much practical knowledge that the experiments could provide them only with marginal assistance. The officials then shifted their attention to organization. They worked out a strategy of linking consumers and marketing cooperative activities. In the process they built support and capacity that could promote improvement in agricultural technology. Farmers in this area were having to pay up to eleven pesos for a kilo of sugar in private stores. Project leaders were able to purchase sugar in bulk from a state organization at a price close to two pesos per kilo. The attraction of a much cheaper price for this staple provided the basis for consumer cooperatives. Project personnel then helped the cooperatives to contract with a state agency that provides other staple goods and canned produce at prices designed to favor low-income people. With the attraction of the substantial savings on essential food and household products, the cooperatives were able to expand their membership rapidly. Project officials then helped the villagers to plan and organize their agricultural production, so that they could bargain more effectively in the market. This movement grew rapidly. The first ten cooperatives were organized largely by staff people, but after that the idea spread and other cooperatives were formed, often having only casual contact with staff members.

Project staff then helped get leaders of the various cooperatives together in regular meetings to discuss and plan their purchases and sales. We attended one of these meetings in 1980 in which representatives of more than forty cooperatives engaged in an active discussion about the chemical composition of the fertilizer they needed, how much was needed in each village at what time, and so on. Project officials explained that before the regional cooperative was organized, small farmers were at the mercy of the state and private organizations producing and selling fertilizer. They had to take whatever chemical composition was most convenient for the companies to supply and to take delivery when it was most convenient for the companies. Now, since they could purchase in large volume, they could specify exactly the composition that they had found most effective for each area and also negotiate the date of delivery for each village. With the organization providing major economies of scale, the farmers in these villages were rapidly increasing their yields and incomes.

The farmers' associations created in Taiwan represent a more comprehensive organizational strategy. Under this system, the rural community becomes in effect a cooperative organization along lines structured by government. The Ministry of Agriculture supports the system by channeling supplies of fertilizer, seeds, tools, machines, and so on through the farmers' association and purchasing their crops through the association. Officially, membership is not compulsory, but the inducements are such that nearly all farm families join (for example, they could not buy fertilizer otherwise). The association has a much broader and stronger base than in the FORUSA-sponsored society, in which membership is entirely voluntary. The buying and selling transactions of the farmers' association provide the association with resources for operating its own physical facilities (warehouse, store, and so on) and for employing at least a manager and an agricultural production professional. Government establishes qualifications for the professional, but the members of the society can vote to appoint him and discharge him. Under this system the extension agent becomes accountable to farmers through their organization (Stavis, 1974).

With practically all farmers in the community as members, the association can manage public works for the benefit of agriculture. For example, the development and improvement of irrigation systems under the FORUSA model necessarily involved negotiating delicate arrangements with nonmembers so that irrigation channels could cross their lands to reach the lands of the members. Under the farmers' association system, the total agricultural area of the community can be treated as a single unit. This larger membership also gives the farmers' association a greater investment capacity than is likely to be found in a comparable agricultural area where cooperative membership is voluntary. Also, of special relevance for our review here, the farmers' associations conduct local trials and experiments to determine the best crops, varieties, and practices for agriculture in their locality.

It is not our purpose to present any ideal model for peasant participation in development. What is best for any country and task depends in part upon the nature of the organization but also upon the culture and social and political relations of that country. Here we are concerned with identifying the general principles involved in the organization and stimulation of peasant participation.

All these cases suggest the importance of changing the usual accountability relationships so that professionals are responsible to farmers as well as to their administrative supervisors. As long as staff in the field are accountable only to their superiors, peasant participation in the process will depend upon the goodwill of those superiors and upon staff skill in stimulating and guiding activities that encourage peasant participation. Given the rapid turnover of higher-level agricultural ministry officials in many countries, a leading official who has promoted a program of peasant participation will often be followed by an official who is indifferent to that objective or actively opposed. In such a situation, government-stimulated peasant participation is bound to wither away.

Accordingly, if agricultural R&D programs are to be the mainspring of accelerated progress involving small farmers, policy makers must make some major changes to give small farmers, as a group, greater control over productive and organizational resources. In this way they become able to hold government staff accountable for performance in the interest of small farmers. This concept should be seen in the broader context of local organizations and local governments generally. Many developing countries have highly centralized government, which may directly appoint important regional and local officials. Even if such officials are elected, their accountability to local residents is not always assured because of socioeconomic biases. In some countries local governments have no taxing power beyond small fees collected for local services. For any major effort, therefore, they are dependent upon funds from the central government. This situation favors the advancement of local officials who can claim to be able to cultivate the friendship and support of officials in the central government and therefore to bring government handouts to their community. Our analysis thus suggests that where such a dependency situation prevails, agricultural and rural development for small farmers will be slow and their incomes will lag far behind those of other segments of the population.[2]

[2]This conclusion is supported by studies conducted by the Cornell Rural Development Committee of local organizations and rural progress in Asian developing countries. They found that in general the countries having the strongest and most active systems of local organization were those which were progressing most rapidly in rural and agricultural development (Uphoff and Esman, 1974). A state-of-the-art paper in preparation by Esman and Uphoff is examining in more detail the organizational dynamics and requisites for involving and assisting rural communities.

Conclusions

As we have seen, a one-on-one relationship between agricultural professionals and small farmers cannot be cost-effective. The conventional research-extension model presents a dual problem. On the one hand, it is inordinately expensive to get technical assistance to small farmers on an individual basis. On the other hand, the difference in status and power between the professional and the small farmer is so great that communication is impaired. In any event, professionals are not responsible to farmers. This situation leads to less conscientious and responsive assistance than is needed to produce widespread agricultural innovation.

Farmer organization can change this situation in basic ways. Working with organizational leaders, the agent can greatly extend the impact of his work. Furthermore, collective organization changes the distribution of power substantially. The small farmer is no longer simply the passive recipient of the initiatives of the professional. He now has some organizational leverage to initiate action with the professional. If that official does not respond, and if other farmers share his views, the small farmer can use his organization to bring pressure to bear upon the superiors of the local staff. This action makes possible (though it does not guarantee) the development of a collaborative relationship between small farmers and government officials.

To be sure, local organization is not a sufficient condition for the progress of small farmers. Government bureaucracies are often so complex and their power centers so far from the local scene that local organizational leaders may be frustrated constantly unless they are able to establish linkages with intermediate power centers that can help them to reach decision makers and to secure coordinated action on the part of the various government agencies that are involved in rural development. In Chapter 13, we examined a case in which a peasant movement in Honduras organized local cooperatives and those cooperatives joined together to form a regional cooperative and, finally, the regional cooperatives joined together to gain a voice in agricultural policy making nationally. We have seen also how technical assistance by government officials played a vital role in helping the cooperative movements to establish coordinating linkages with government at regional and national levels. Such an organizational evolution greatly strengthens the capability of a national system of agricultural research and development.

15

New Directions in Irrigation Research and Development

The performance of an irrigation system depends upon a combination of physical and socioeconomic factors. Such systems generally involve both local groups and national or state government agencies. Irrigation is therefore an ideal topic for the exploration of the interrelations and interdependency of the various organizational levels on which we focus in Part IV.

Interdependency of Main System and Local Units

Groups of farmers often divert water from a river and operate an irrigation network to supply their own farms independent of any government agency. In some instances, a government agency may regulate the amount of water removed from the stream and assist in resolving disputes that arise between upstream and downstream users.

Where irrigation is based on a single large structure diverting water in an expansive area of irrigated farms, government is necessarily involved in controlling the water supply and establishing the terms under which user communities can receive water from the main system. Theoretically, government control could extend from the source to the last drop of water used on the last farm in the command area (the area irrigated by the system), with government employees operating the turnouts and determining the timing and quantities of water flow throughout the area. Such an arrangement would be impractical for several reasons: the exorbitant costs of government administration, the near impossibility of the agency acquiring the location-specific information required, and the high level of distrust regarding agency fairness and efficiency.

Recognition of these problems has led to the widespread development of irrigation systems in which responsibility for administration and maintenance is shared between government agencies and local irrigation organizations.

Recognition that responsibility should be shared does not tell us anything about the distribution of labor and authority between government and local user associations. Neither does it tell us how the costs of managing the system are to be shared or collected.

Until recently, government planners have concentrated their attention on the physical and economic aspects of big projects such as the construction of a dam and have given little attention to the nature and problems of managing the system. We are now coming to realize that major management problems exist both at the level of the main system, where agencies usually are involved, and at the local unit level, where farmers are involved.

Operating an irrigation system requires organizational policies and activities to accomplish three fundamental purposes: water allocation and distribution, physical maintenance of the system, and conflict management.

Controlling the flow of water so that each farmer is equitably served requires one or more individuals with at least part-time assignments to manage the intake from the main system and to distribute the flow to the various fields. Such individuals must be compensated in some way: through payment in cash or kind, gaining use rights to certain lands, being free of other maintenance obligations, and so on. Most local systems need major maintenance work at least annually to clean the canals, to eliminate obstructions, and to make repairs. If this is a collective responsibility of all farmers, the users must devise a system for mobilizing and directing their own labor. Except in rare cases with a surplus of water available for all users at all times of the year, there are bound to be conflicts, some users claiming that they are not getting the water due them because those controlling allocation are favoring their own farms or those of their relatives and friends. Any system calling for communal work in physical maintenance is bound to encounter instances of individual farmers appearing not to be meeting their obligations. Unless the user organization can resolve such conflicts and impose penalties on those failing to meet their community obligations, the efficiency of the local system will be jeopardized.

In this chapter we deal with some of the operational and organizational problems observed in recent irrigation research and development projects, primarily in South and Southeast Asia. During the 1960s and 1970s, these regions have had high levels of actual and planned irrigation investment (Oram et al., 1979; Colombo, Johnson, and Shishido, 1977). Most observers agree that irrigation development will be a major component of agricultural change in Asia and the Middle East for the next several decades.

Common Operational Problems

Increasing evidence suggests fundamental and recurring operational problems with effects on ecology, production, and equity. Among the unexpected outcomes, the following have been reported: (1) problems of waterlogging and salinity of soils within and below irrigated areas resulting from water leakage from conveyance channels and overuse of water in farm fields; (2) the uneven spread of modern rice varieties because of unsatisfactory water control and unreliable delivery in some irrigated areas; (3) low irrigation efficiency (more water supplied per unit of land than is required for normal plant growth) resulting from poor design, operation, or maintenance of the irrigation apparatus; insufficient leveling of land; or farmer ignorance or inappropriate incentives; (4) underutilization of irrigation, which may be related to uncertain water delivery schedules or other factors such as inadequate labor supplies or unappealing prices; (5) slow development of terminal-level channels and ditches to move water to farms from the main system; and (6) low level of irrigation fee payment and limited investment by users in maintenance. Recognition of these problems has induced a number of efforts to gain a greater understanding of the problems and their causes and efforts to implement irrigation development in new ways.

Irrigation R&D in the 1960s and 1970s took on several important characteristics not found in the earlier work. Four of the more significant were emphasis on field data collection on actual operating irrigation systems; use of a broad multidisciplinary mix of perspectives in research, including problem identification, data collection, and analysis; close working relationships with the agencies responsible for implementing irrigation development; and modification of some components of the system being studied (often part of the physical apparatus, sometimes management procedures) and evaluation of results.

Field Research in Public Irrigation Systems

One of the first field-based research activities involving an interdisciplinary team of scientists was the Colorado State University (CSU) project in Pakistan implemented in cooperation with the Water and Power Development Authority. Much of the research was done in the Mona Reclamation Experimental Project in Punjab province. A team of CSU and Mona scientists collaborated in assessing actual water flows and farmer activities and in identifying and testing various technological innovations (Colorado State University Water Management Field Party and Mona Reclamation Project Staff, 1977).

The CSU work concentrated on terminal-level watercourse units covering an average of 160 hectares. Water is delivered from the main system through a fixed turnout structure and farmers are expected to operate and maintain the watercourse facilities. Researchers observed that high water losses occurred in watercourse commands. These losses and their effects in reducing irrigated acreage and crop yields also contributed to waterlogging and salinization. Having thus defined the problem, the researchers sought solutions in new technologies to line the watercourse channels, better structures to regulate diversion from the watercourse channel to individual fields, and improved farmer organization and knowledge to manage the watercourse unit.

One result of the project was the decision by the government of Pakistan and various external agencies to invest in the physical rehabilitation and improvement of watercourse facilities throughout the country. These construction activities were to be complemented by efforts to create watercourse irrigation groups to maintain and use the facilities. Regrettably, the CSU project was unable to do organizational research, and the lack of research-based guidelines in combination with a largely uninterested bureaucracy has resulted in little attention to these key matters.

It is now obvious that, in concentrating at the watercourse level, the CSU work failed to examine important issues about management of the main system. But the CSU project should be recognized as a critical pioneering effort in establishing the need to conduct irrigation research on site with a multidisciplinary perspective. Unlike many studies previously conducted in Pakistan, this research was done neither in the hydraulic laboratory nor on the experimental farm.

In the early 1970s, the Irrigation and Water Management Department of the International Rice Research Institute (IRRI) began research with the National Irrigation Administration (NIA) in the Philippines. This project sought to "develop, implement and evaluate a package of management practices for improved distribution of water along a lateral canal" serving approximately 5,700 hectares (Tabbal and Wickham, 1978: 94).

During the first year, field data were collected to measure water flows, stress days, and crop yields associated with the existing lateral operation. During the second year, a new system of water distribution was implemented and similar data collected. The new distribution arrangements included two innovative procedures: the assignment of priority to the lower half of the lateral's command area during the wet-season transplanting period (thus reducing the total transplanting time for the entire lateral command in the wet season and advancing the date for dry-season transplanting); and a more controlled distribution of water to the sublaterals based on assessment of needs and water availability.

Research indicated several remarkable outcomes of the modified water management procedures. One significant result was that farmers in the lower half of the area (the so-called "tail enders") were able to increase the amount of land planted to rice in the dry season. This increase, along with improved delivery throughout the area which served to increase average yields, raised the total production.

In addition to these immediate practical results, this research was influential in demonstrating that the water management problem was the result, in part, of agency actions in operating the canals and not simply of poor water use by farmers. Although now a more widely accepted idea, at the time most policy makers (and many researchers) viewed poor system performance as the result of inadequate farmer knowledge and inappropriate farmer behavior. As a result of the IRRI research and subsequent work, there is now much more concern with improving the performance of the main system (the headworks, canals, and other structures administered by the irrigation agency) as well as that of the tertiary-level, or farmer-managed, components. It is now more commonly assumed that some of the "irrational" behavior of farmers observed in irrigation systems may be a form of adaptation to the uncertain water environment created by the irrigation bureaucracy.

R&D with Community Irrigation Systems

Separate from the research team efforts just discussed, social science–oriented research in the 1970s focused on small-scale community irrigation systems, which had been largely ignored by the irrigation agencies and engineering researchers. This research was also motivated by a desire to observe and understand existing irrigation units. Unlike the social science research that preceded it, the research of the 1970s was both more micro-focused and more development-oriented.

The research advanced rapidly in the Philippines during the last decade for several reasons, including the fundamental importance of community irrigation (serving approximately 50 percent of the total irrigated area of the country), the irrigation agency's past involvement with such systems and its interest in improving assistance to them, the existence of previous research on these systems suggesting new approaches, and the availability of extremely able Filipino researchers.

A central element was the work done by staff of the Institute of Philippine Culture (IPC). Under contract with the National Irrigation

Administration, IPC selected a sample of community irrigation systems for study (de los Reyes, 1980b). The general objective was to provide the NIA additional information on fundamental characteristics of these systems so it could devise improved programs of assistance. Special attention was given to the organizational forms and management procedures used in these systems as well as to their technical features.

Among the important consequences of the nationwide research was that it helped to moderate common misconceptions regarding the inefficiency and ineffectiveness of community irrigation. Researchers concluded that the variety of approaches used in community systems "attest to the farmers' ingenuity and capacity for meeting the demands of irrigation management. This belies the popular belief that farmers lack the know-how to operate and maintain their system" (de los Reyes, 1980a: 6). The picture presented in this study was of a community irrigation sector in which considerable local investments were made regularly and which had substantial infrastructure and institutional resources. Not everything worked perfectly, but a combination of appropriate technology and organization often resulted in improved system performance. Given the breadth of coverage provided in the study design (forty-seven community systems were surveyed in five selected regions of the country and four detailed case studies were conducted), the results were not unrepresentative or inconsequential.

In addition to these important facts about community irrigation systems, the study also developed several individuals with unique information about and interest in these previously little-known systems. This combination of IPC facts and staff was an important product of the research, consistent with the ideal R&D pattern, and became important to the actions discussed below.

The NIA had formerly provided occasional assistance to community systems (Bagadion and Korten, 1979). In 1974, however, it was given a more specific mandate to assist communal irrigation and, as with the large public systems, to recover from the beneficiaries a portion of the costs. Thus the agency was interested in the IPC irrigation survey as background for understanding the communal section and was motivated to translate this and other information into better strategies and approaches for working with these systems—in particular, to develop new procedures that would increase the possibility of repayment by the assisted groups.

The initial search for new approaches was concentrated in the Laur project (Isles and Collado, 1979). The approach related to two elements of the then existing situation: the emerging evidence that farmers in many communal systems had considerable expertise and experience in operation and the NIA mandate to recover a portion of the assistance costs from the community. The latter point was especially important because it suggested a redefinition and reassessment of agency performance. Not only was the agency to build new irrigation structures or expand the service area of a particular community system, but it was also to obtain repayment from the community for a portion of these costs. The NIA recognized that repayment would be made only if the community was satisfied with what it had received and if it could be held legally responsible for the debt. From this realization emerged the central idea of the Laur approach "Since it was the farmers who would be expected to pay for, operate and maintain the system once construction was completed, many felt that they should be heavily involved in planning and constructing the system. This prompted the idea of initiating a project in which maximizing such farmer involvement was an explicit objective" (Bagadion and Korten, 1979:7).

Two fundamental elements of the initial strategy were the recruitment of specialized staff to assist the development of local organizations (called community organizers) and the rearrangement of the usual irrigation development sequence to let local organizations take the first steps in design and construction activities. This strategy was intended to allow farmers to improve their organizational capacity, consider their present situation, and formulate an agenda before dealing with the assistance agency. It was assumed that this preparatory organizational work would increase the exchange of experiences and knowledge between the bureaucracy and the community and strengthen community commitment to any subsequent irrigation development.

Lessons Learned

From this initial effort and later expansions, a number of important irrigation development lessons have emerged. The more important are discussed below.

Design is not a task for professionals alone. Where farmers already have experience with an existing irrigation system, experience with and knowledge about its past performance, the physical environment, and the social context are all important to any improved design. Outside professionals will have the difficult, if not impossible, task of trying to incorporate this information unless farmers are included in the design process. The idea is not that the irrigation association is to be provided with survey equipment and drawing boards and expected to prepare technical specifications for structures and canal layouts. But villagers can have important inputs into such fundamental design issues as the area to be served, the cropping pattern to be followed, the basic pattern of water distribution, the identification of particular problem spots and zones of special need, the existence of historical precedents, and important social boundaries. All of this knowledge can result in a richer map of the existing setting from which a design can be prepared by professional designers for farmer reaction.

This interactive process holds the potential for preparing designs that are both technically superior and socially appropriate.

Considerable local resources can be mobilized in support of irrigation development. Community organizers mobilized a number of resources, the most fundamental being local organization. Because of the initial emphasis on strengthening the local irrigation organization, by the time the design and construction phases began, this important resource was already operating. The existence of this local organizational capacity allowed the systematic use of knowledge, as in the design process discussed above, and of labor during the period of construction. Requiring farmers to repay a portion of the project costs provided an important incentive for them to keep costs down by offering their own labor. But careful planning, information dissemination, and recordkeeping by the irrigation organization were needed to arrange for the right number of people to arrive at the right time and place to get jobs completed. Finally, as the payment of irrigation fees begins, these resources will be used in expanding the agency's work with other communal systems.

Modifications in the agency's structure and procedures are required to sustain farmer involvement. Commonly, irrigation agencies are not organized to operate with high farmer involvement. The culture of professional engineering and the usual standards and procedures for rewarding staff, controlling budgets, and assessing performance all serve to concentrate decisions and implementation in the agency professionals. If greater farmer involvement in design, construction, and operation is to spread beyond specific pilot projects, changes must occur in the agency's approach. This is perhaps the most important lesson from the Philippine experience: expanding farmer participation in irrigation development requires not only better farmer organization but new agency structures and processes.

Budgeting is one illustration of the lack of fit between a participatory strategy and the usual agency procedures. For example, the National Irrigation Administration, like many government agencies, is required to return to the national treasury funds which it is unable to disburse during a particular year. "This places pressures on engineers—and farmers—to avoid any activity such as participation which might extend the construction period. One of the very clear lessons of Laur is that building social capacity takes time and schedules must be adjusted accordingly" (Alfonso, 1981:51).

Even when professional staff value the involvement of farmers in designing and implementing irrigation development, they may be constrained by the structures and procedures of their agency. One approach to resolving these structural problems has been the NIA's creation of a Communal Irrigation Committee. This committee is composed of key NIA administrators and selected outsiders who have expertise with communal irrigation development, including staff from the Institute of Philippine Culture, the Asian Institute of Management, the International Rice Research Institute, and the Ford Foundation. This group meets with staff of various communal projects to review and plan needed research and training.

Improving Large-Scale Systems

The initial successes of this work with community systems has prompted interest in the applicability of these approaches to the development of large-scale irrigation schemes. Two such efforts are now under way—the Buhi-Lalo project in the Bicol region of the Philippines and the Gal Oya project in Sri Lanka. Our discussion will deal with the latter project.

The Gal Oya project in the eastern dry zone of Sri Lanka has been operating for about twenty-five years. In 1980, with assistance from USAID, the Department of Irrigation initiated a plan to rehabilitate the physical facilities of the Left Bank portion of the project (an area originally designed to irrigate approximately 15,000 hectares, but now estimated to be irrigating 24,000–26,000) and to improve water distribution and use in this zone. Management is to be improved by creation of water user associations and additional training for the agency staff.

These project objectives and the stated means for achieving them are not uncommon, but this project includes an important difference: "What is significant is the attempt to depart from a typical engineer-planned-and-implemented reconstruction effort by consciously drawing upon the nearly twenty-five years of farmer experience with the Left Bank system. Active farmer participation in the project is expected at both the design and reconstruction stages and, following completion of physical rehabilitation, in the management and maintenance of the system, at least at the tertiary or field channel level" (Vander Velde et al., 1981:2–3).

As with the Philippine experiences previously discussed, achieving active farmer participation in the design and construction stages is a complex task and must deal with both agency and community structures and processes. The responsibility for designing and testing strategies to achieve farmer participation has been assumed by the Agrarian Research and Training Institute, an experienced Sri Lankan group that operates with government support and with staff support from Cornell University.

In designing an experimental approach to participation in Gal Oya, the Philippine experiences were reviewed and selected elements adapted. For example, institutional organizers were recruited and trained for a pilot area of 2,000 hectares (each organizer covering an area of about 50 hectares). The organizers enter the field before design and construction activities have begun (though the initial assignments provided only a short gap between arriving in the field and initial farmer-agency discussions). Important early tasks of the organizers include identifying turnout boundaries and the farmers included in them, beginning a discussion with the farmers about their irrigation problems, and assisting in organizing the initial meetings between farmers and the Irrigation Department staff.

There is an important difference between Gal Oya and the community systems of the Philippines, some of the implications of which emerged early in the project. The fundamental difference is in governance. In the Philippines the system is owned and operated by the community, and assistance is provided by the irrigation agency. In contrast, the Gal Oya system belongs to the government and is governed (in theory) by the Irrigation Department. From the viewpoint of this agency, farmer groups are to assist the agency in carrying out its responsibilities. The negotiation of rules and procedures for farmer-agency interactions, as well as joint and separate responsibilities, thus proceed from a very different base.

Some of the emerging problems are similar to those with community irrigation development. For example, timing and project schedules are sources of stress. Even though serious delays in the project may result from factors unrelated to farmer participation (delays in arrival of equipment, slowness in assigning agency staff, or the lack of basic maps for planning), there remains the concern that having to involve farmers will result in prolonged delays.

The concern with timing may simply be a symptom of more fundamental agency concerns with losing control. As has been noted, the Irrigation Department's view that it is the controller and manager of water in Gal Oya makes it uneasy in defining the responsibilities of water user associations.

> Early in the project, much attention was given to identifying the powers and responsibilities of the irrigation associations that were to be formed in Gal Oya. However, the discussions then and the agreements with the Irrigation Department that followed tended to be too abstract, too divorced from the reality of farmer participation and organization development that had begun to emerge several months later in the project area. It is now evident that the Irrigation Department didn't appreciate, or perhaps didn't fully understand the implications of its earlier agreement to give farmer organizations control *at* the field channel turnout gate, and even share in irrigation operations and management at higher levels in the system if turnout groups merged into a higher order federation. (Vander Velde et al., 1981:15)

The Irrigation Department is not to blame for its uneasiness. The department is the victim of unclear national policies that fail to state the desired mix of agency and community responsibility or fail to establish new forms of accountability. In this undefined situation, it is understandable that the Irrigation Department is reluctant to transfer formally a part of system governance to water user groups (the de facto transfer that has occurred can, of course, always be attributed to problems such as deteriorated facilities, inadequate staff, or uncooperative behavior by farmers).

India's Command Area Development Program

Nearly half of India's irrigated area is served by large-scale government canal systems (approximating a staggering 25 million hectares). In the mid-1970s, in response to the widely held perception that these systems were performing very ineffectively, the Indian government initiated its Command Area Development Program (CAD), a complex program that was intended to work at two levels. At the top a unified project agency was to be formed that would administratively control the relevant agencies (usually including irrigation, agriculture, cooperatives, soil conservation, and revenue). Irrigation development activities at the on-farm level were to include four principal tasks: the extension of field channels from the last government outlet structure to farm fields; land development (primarily leveling of individual fields) and consolidation of scattered plots; initiating a fixed rotation (the so-called warabandi system) for the allocation of water to farmers in a given turnout unit (typically twenty to one hundred hectares with ten to fifteen owners); and the organization of formal water user associations at the turnout level (Wade, 1980).

The innovations emerging from the CAD program have not come from work at the on-farm level, which needs improvement. Some of the fundamental program components seen in the Philippine and Sri Lankan experiences have not yet been tried in the Indian project. For example, there has been no use of specialized field workers (community or institutional organizers) to help establish water user associations, and it appears that local groups exist largely in name only. Because of this lack of careful attention to local organization, farmers have not become involved in the design of the new on-farm level facilities. Thus the designs have not incorporated their knowledge nor has the design process been used as a means to inaugurate new farmer associations.

But, relating to what Wade and Chambers (1980) have referred to as a blind spot, from the CAD program a clearer understanding has emerged that the main system (the portion administered by the professional irrigation staff) frequently is operating very poorly and that this inadequate performance has large implications for the poor performance of the system at the farm level.

In at least one case, these new insights arose because a nonengineer was responsible for the administration of a large irrigation project. In the Pochampad system in Andra Pradesh, a senior agricultural extensionist was made head of the project authority. In planning implementation of tertiary-level improvements in a portion of the service area, he sought assurances from his irrigation staff that water could be reliably delivered to the zone involved. He feared that he would be unable to sustain farmer interest in improving the tertiary units if they received inadequate water supplies. To his great dismay, he was informed by the engineers that they could not supply adequate water.

As Hashim Ali (former head of all Command Area Development programs in Andra Pradesh) has stated, the poor performance of the main systems in his state is related to three fundamental problems (Ali, 1980).

1. Outmoded engineering design: canals and other structures designed to spread water supplies widely in times of drought to avert famine are now being used to serve the more demanding needs of modern wheat and rice varieties.

2. Insufficient maintenance funds: governments commonly have assigned meager funds to maintenance, and the deteriorated facilities are incapable of delivering sufficient and reliable supplies.

3. System operation is professionally unimportant. This task has

never been as rewarding for engineers as the design and construction of impressive new physical structures.

To this list, one might add that thus far farmers are only peripherally involved in the state's irrigation development activities.

Conclusion

In Asia, irrigation development will remain a decisive component of agricultural development for the next several decades. But the actual contribution of irrigation to high and sustained crop production will be limited unless new R&D procedures are developed and applied. Some new procedures are beginning to emerge. As illustrated above, though they are significant, they are limited in scope.

As irrigation research in the 1970s increasingly turned to the field, our understanding of what is actually happening in irrigation systems has increased. Likewise, our ideas about how to solve contemporary problems have been changing. For example, the initial calls by policy makers and project implementers for the creation of water user associations stemmed more from their frustration with system performance and their preconceived ideas of what farmers should be doing than from an understanding of what farmers already were doing and were capable of. Actual observations of farmer (and agency) behavior in distributing water, maintaining the irrigation structures, and managing conflict have changed our ideas about the timing and extent of farmer involvement in irrigation development activities.

As with other programs of agricultural development, many now believe that it is vital to bring farmers into the irrigation development process. In addition, it is more widely recognized that irrigation development means better water management by the irrigation agency as well as by the water users and that both increasing the involvement of farmers and improving the agency's water management activities require new and retrained staff. Most irrigation agencies do not have staff trained in assisting the formation of water user organizations. Yet if creating organizations is an essential part of the irrigation development strategy, such specialists are essential. In addition, considerable effort will be needed to train the regular engineering staff in a variety of new skills, including both improved water management at the system level and new skills to incorporate farmers into their design, construction, and operations activities.

Efforts are under way to implement irrigation development programs that have these characteristics. The monitoring of these efforts and the assessment of results deserve our attention and review, as does a continuing concern with expanding the capacity of groups elsewhere to design and implement such innovative approaches.

16

Bureaucratic Reorientation for Agricultural Research and Development

Integration of small farmers into the process of technological and scientific progress depends upon pronounced changes in the relations between professionals and peasants so that peasants become active participants in the change process. We have shown how this may be done at the level of the village and in the farmers' fields. If the emerging strategy for rural development is to advance beyond pilot projects, however, the new field-level strategy must be linked to, and be supplemented by, a fundamental reorientation of agricultural bureaucracies. If governments retain the existing structures and processes of agricultural bureaucracies, it will be impossible to implement a participatory model in the field. Therefore, we must consider the nature of the bureaucratic reorientation required for carrying out such a participatory strategy in rural communities.

In this chapter, we will consider first the political and then the bureaucratic barriers to change in traditional government agricultural agencies. We will then conclude with an analysis of the change process involved in reorienting bureaucratic organizations so as to provide support and guidance to programs built upon the active participation of small farmers in research and extension.

Barriers to Change in Traditional Agricultural Bureaucracies

Those seeking to build a participatory R&D system are faced with both political and bureaucratic barriers to change. "Political" here refers to the policies and activities of the decision makers, such as the minister of agriculture and his immediate superiors and subordinates. "Bureaucratic" refers to the patterns of behavior we can expect to find in the day-to-day operations of a large organization. While the two categories obviously overlap to some extent, let us start with the political because, unless changes are introduced here, there will be little possibility of reorienting the agricultural bureaucracy.

In developing countries, as in industrialized nations, personal connections can often be more important in securing a high administrative job than demonstrated performance in the field. Therefore, even if the leader of the organization has gained his position largely through performance and is committed to reorientation of a bureaucracy, he generally has trouble motivating subordinates whose capacities are limited. Furthermore, since the bureaucratic organization implicitly serves the purpose of providing employment as well as carrying out its official mission, the leader will not find it easy to transfer or discharge employees he considers ineffective or irresponsible. Even when employment is not threatened, the job interests of an organized employee group may present a major barrier to reorientation. For example, let us say that the leader would like to carry out a major reorganization of the duties and responsibilities of agricultural extension agents but finds, as in some countries, that the agents are well organized politically so that his efforts to transfer them or otherwise change their duties and the location of their activities meet political resistance.

The instability of political positions in agriculture can also be a major obstacle. For example, in one nation there were fifteen ministers of agriculture in a recent twelve-year period. Few people who attain such a prominent position are content simply to administer programs turned over to them by their predecessors. The new minister naturally wants to have some personal impact upon agricultural development. Since he knows that he will not have many months to achieve this impact, he prefers to inaugurate projects that will have maximum public visibility. If he approves a project to begin building a dam to irrigate previously arid areas or to start a new colonization project, he can dramatize the event by appearing at the site and making a speech to workers and community people, while his aides assure maximum media coverage. The big project will not yield any benefits until months or years after he has left his post, but it is easy for him to dramatize his desire to assist the community by launching such an ambitious project. And if the project eventually turns out to be a fiasco, as happens to many grandiose projects, he can argue that the failure was the result of mismanagement by his successors. They in turn can blame him for a poorly conceived project.

The new direction argued for here points administrators away from the big and spectacular projects that are easy to dramatize. It is much more difficult to dramatize the reorientation of bureaucratic structure and processes to develop a more participatory system of relations with small farmers. Even at the level of the agricultural experiment station, the new direction emphasizes less elaborate physical structures and a shift of activities off the station onto the fields of small farmers. There the results under actual field conditions are bound to appear less impressive than those achieved under the favored conditions of the experiment station. Furthermore, it takes considerable time to work through a reorientation of the agricultural bureaucracy and still longer before the results show up in a widespread improvement in the welfare of farm families and rural communities.

There is also a major difference in the level of understanding required to lead a traditional organization through a basic reorientation process. The leader who plans no major changes in organization structure can carry out his duties without public embarrassment even if he has little understanding of what is going on because he is surrounded by associates and subordinates who are accustomed to the traditional ways of operating. A leader who aims to bring about a major bureaucratic reorientation must have a much higher level of understanding both of the traditional bureaucracy and of the process through which he hopes to bring about change, as well as the organizational form that would accomplish his objectives. Such a leader must have far greater skills in articulating his objectives and presenting the rationale for his decisions than are required of the traditional leader. Unless subordinates and associates see a particular move as part of a general pattern, they will not understand the rationale behind it and will therefore be ineffective in carrying it out. It is the responsibility of the change leader to make clear both the general pattern that guides him and the place of the particular move within this general pattern.

An effective change leader will appreciate the pitfalls involved in trying to avoid bureaucratic frustrations by seeking shortcuts that enable him to advance particular projects outside of the existing framework. Pilot projects represent one such apparently convenient shortcut, which avoids traditional inefficiencies and blockages by going outside the bureaucracy altogether to start the new project. Even when staff members are government employees, which they usually are, they have exceptional support and autonomy. If getting the job done means simply increasing yields of a particular crop in a particular small area at a particular time, this strategy can often produce results.

According to the usual rhetoric, the pilot project is to serve a broader and more far-reaching purpose: that of guiding the agricultural bureaucracy toward more effective and efficient operations in the future. Unfortunately, project planners or organizational leaders almost never give much thought at the outset to the ways in which this new manner of working can be integrated into the national program so as to have a constructive effect upon the total agricultural bureaucracy. They have faith that success in a small project will naturally lead to the diffusion of ideas and social processes into a larger arena. On the contrary, more often than not we find that the pilot project functions as a small enclave within the territory of the bureaucracy, having little effect upon it. That this does not necessarily have to be the case is shown by the Caqueza project, but in this case project planners and organizational leaders devoted considerable thought and discussion at the outset to the relation between the pilot project and the agricultural bureaucracy and continued to work over the problems of relating the project to the bureaucracy throughout the course of the project.

Indonesia provides a classic example of a fiasco that can result from efforts to bypass the existing bureaucracy. As we have written earlier of the Bimas case:

> Given the urgency of the food production and farm income problems, politicians are inclined to pick up any plausible new idea and seek to mount a program to put that idea to use on a nationwide basis or at least in some large region. Perhaps the most spectacular example of this tendency is what came to be known as the Bimas case in Indonesia. In order to attain self-sufficiency in rice, government decision-makers in 1968 established a goal of increasing production by somewhat more than 50 percent within a five-year period. Recognizing that such an ambitious objective was beyond the capacity of the government, the planners decided to by-pass their own agricultural bureaucracy by channeling the flow of new and increased inputs through commercial organizations. Still, the agricultural bureaucracy could not be left out altogether, since regional administrators were needed to set regional production targets and report on results. After less than two years of confusion and conflict among government agencies and private firms involved in Bimas, the whole program collapsed. As Hansen notes, "The size of the target overwhelmed the existing structures of administration and substantially reduced the effectiveness of the entire campaign to achieve self-sufficiency" (Hansen, 1972). [Whyte, 1975:45]

Another common failing in organizational change programs is that they often attempt to short-cut stages of development. When top government officials believe that a particular project has achieved success, they are naturally inclined to try to jump in one step from the small area to the nation as a whole, or at least to a large region. As we have seen in the Comilla case, such sudden and rapid expansion resulted in a substantial loss of effectiveness. A still more clear-cut example of failure through too rapid expansion is provided by the case of an AID training program to improve the abilities of extension agents in Pakistan (Green, 1961). The program was innovative, emphasizing group discussion, field work, and practice demonstrations of the skills agents were trying to transmit to farmers. Extension agents responded with enthusiasm, and government officials visiting the training center were so impressed that they decided that the program must be extended at once throughout Pakistan. Members of the training team protested that they had not had time to consolidate the gains and should therefore delay expansion and then proceed in gradual stages.

Government officials insisted that what was good at this local training institute must be implemented in all similar schools throughout the country, and the training team spent the next year on the road, conducting successful programs in one center after another. When they finally returned to the original training center, they found no trace of their innovative program. The institution had reverted to its traditional ways of organizing time, professors, and students. Thus the sudden expansion resulted in the withering of the original program in the base center, without permanently implanting that program in any other training centers.

In starting our discussion with problems imposed by political leadership of agricultural bureaucracies, we do not mean to give the impression that freeing the professionals from the interference of the politicians would solve all problems. On the contrary, we find major deficiencies in the orientation and conventional style of agricultural professionals. Generally they do their planning at a site remote from where the plans are to be implemented and receive little if any input from the villagers for whose benefit the plans are presumably being made. In developing nations, the planners are always far above the small farmers in social class, and they may also be separated by ethnic and cultural differences. In some countries many small farmers may speak indigenous languages that are incomprehensible to the planners.

If they get into the field, all too often the planners do so only on two occasions: when they make an initial "windshield survey" of the area to be affected and again when they come to persuade local people that the plan will be just what they need. When this top-down approach does not yield the benefits the plan calls for, the planners fall back on the conventional excuse that we have called the myth of the passive peasant. There was nothing wrong with the plan, but the peasants were locked into their traditional culture and resistant to change.

If planners and administrators are to cope with the obstacles to successful change programs described above, they must begin by developing a better understanding of the nature of traditional bureaucracies. We will discuss bureaucracies in general and then focus upon agricultural bureaucracies.

Researchers studying large bureaucracies have found a common pattern. Initiation of action upward tends to be difficult. Communication tends to be heavily filtered because good news rises rapidly and bad news tends to be held back and passed upward only in modified and distorted form, if at all. Subordinates tend to concentrate on trying to please their superiors, on whom their job security and advancement depend, even at the expense of the clients they are supposed to be serving. Their emphasis is often on maintaining good relations within the office, where bureaucratic subordinates have opportunities to interact with their superiors, even if they have to reduce contacts with clients, especially clients of low status and without political power. This tendency is likely to be accentuated in the agricultural ministry of a developing country, where there is a large gap between officials and small farmers. This social gap is too large to be bridged by informal adjustments. We need to develop structural arrangements that will make the extension service more open and receptive to initiatives from small farmers, who lack education, status, and influence.

In the traditional bureaucracy, orders and instructions may go from the top to lower levels with reasonable speed, but achieving compliance is another matter. Defiance of orders is generally rare, but evasion is common. This problem tends to be accentuated in organzations whose officials are spread out over a wide territory, as is typically the case with extension services. The problem is further complicated by the differences in culture and social life between the capital, which in many developing countries is the primary city, and provincial cities and towns, which are likely to be looked upon by officials assigned there as what we call the boondocks. Since superiors do not get much pleasure from spending time in the boondocks, their visits to the field may be too infrequent and too brief to enable them to gain a real understanding of the problems of the local agents.

The problems of leadership and control are complicated by the difficulties in developing adequate procedures and standards for judging the performance of agents in the field. To some extent these evaluations can be put on a quantitative basis, but that method can produce undesired results. For example, suppose the agent is to be judged on the amount of farmland in his area where a new high-yielding variety of plant has been introduced. To oversimplify a common situation, let us say that his area has one farmer who owns 100 hectares and a hundred farmers owning 1 hectare each. Under this quantitative evaluation standard, the extension agent is naturally inclined to concentrate on the hundred-hectare farmer. If he adopts the innovation, the figures will look very good immediately. Furthermore, the hundred-hectare farmer is likely to have more in common socially with the agent, and he will have more resources that he can afford to risk than the one-hectare farmers. In such a case, government policy to maximize the amount of farmland that is devoted to a new variety will have action implications that directly violate another stated policy calling for special attention and help to small farmers.

A program based upon top-down pressure to achieve quantitative targets can lead to simulated compliance in which subordinates falsely report the results achieved. When this happens, decision makers increasingly base their policies and plans on misinformation. For example, in Tamil Nadu state in India a careful study demonstrated that only one-third as many hectares had actually been planted in high-yielding varieties of rice as had been officially reported (Farmer, 1977:96).

The traditional bureaucratic structure tends to lead people to vertical relations of power and prestige. Each functionary tries to expand his own turf or at least resist the incursions of rivals. This emphasis upon vertical relations and power makes coordination of activities horizontally exceedingly difficult.

Conventional Value Orientation

Four value equations, reflecting deeply held assumptions regarding the nature of the development process, tend to reinforce prevailing bureaucratic practices and the procedures and reward systems that support them. Since these values rationalize bureaucratic convenience and advantage, even though they pose barriers to effective development performance, they are seldom questioned by administrators and constitute stable views of the world. These presumptions are deeply institutionalized in bureaucratic practices and routines. Yet introducing alternative views will be important, even essential, to the process of bureaucratic reorientation.

Expenditure = Results. For years, prevailing economic theory has equated development with growth of GNP and has attributed such growth to increases in the level and rate of capital investment, assuming that labor and natural resources are freely available with no opportunity cost in foregoing other options. This view has come under increasing attack, and few would advance it now, at least in its earlier pristine form.

But bureaucratic practice has reflected a formulation that has received less critical attention: *expenditure = results.* The preoccupation of government bureaus and donor agencies with "moving money" and their predisposition to take as the primary measure of development performance the rates at which their staff is able to spend development funds treats such expenditures as a proxy for development progress. Indeed, even donor countries now find their development contribution being assessed in international forums in terms of development resources transferred, as a percentage of GNP. Those who move the most money are the heroes, whether at the level of international assemblies or among the lowest bureaucratic functionaries. Those who do not move increasing amounts of money at ever faster rates are dismissed as laggards, without consideration of the comparative results of expenditure—how much, for whom, and lasting for how long.

Even though experience suggests there is little association between the size of government expenditure and the results achieved, administrators and technicians are under inexorable pressures to act as if there were such an equivalence—to design programs and implement them with all possible speed so that they and their superiors can report X amount of progress in terms of Y amount of money spent. Such pressures are detrimental to poverty-focused rural development, because participation and the development of new bureaucratic capabilities necessarily take time, and any activity likely to introduce delays in project approval and implementation is not viewed kindly.

The incipient movement within the development profession to institutionalize evaluation in connection with programs and projects is a well-intentioned effort to create a countervailing accountability for real results. But generally it identifies problems only after the fact and has little influence on decisions or on careers. Too often evaluation, if done, is done with reference to the initial conception of the problem—how well and timely did the organization do what it set out to do—rather than dealing with solving of priority problems as they became more evident from the accumulating experience.

A more participatory approach to development is unlikely to develop until the implicit equation of expenditure with development progress is removed from our thinking and practices, from our evaluation criteria, and from our budgeting and authorization systems. A major effort and time will be required, but a strategy of bureaucratic reorientation needs to have long-term as well as short-term elements.

Education = Superior Wisdom and Virtue. Perhaps even more deeply ingrained is the idea that advanced formal education, symbolized by higher degrees, makes the possessor generally superior to persons with less education. Too often the presumption is made, by expert and nonexpert alike, that the judgment and even the values of the educated person should prevail. This view truncates participatory processes (Conlin, 1974; Chambers, 1979).

The knowledge of well-educated persons is usually necessary but not sufficient for program planning and implementation. There is much local knowledge which the expert is unlikely and sometimes even unable to acquire, which local people can and should contribute to decision-making processes. Local knowledge is unlikely to be sufficient, but it is almost always necessary for efficient use of development resources, especially when the objective is to assist the poor majority.

One of the purposes of getting more popular participation in development is to achieve a fruitful combination of expert and indigenous knowledge. No useful purpose is served by humiliating experts (they will not become more receptive to local inputs), but it is necessary for them to become more humble. For example, in the Philippines and Nepal, local people told engineers that the dams they were planning to construct would not be strong enough to contain the force of the rivers at flood. The engineers insisted they were correct, demonstrating the efficacy of their designs and materials with mathematical formulas and references to successful structures elsewhere. Some months later, the dams were washed out, demonstrating that persons knowing local conditions, however lacking in formal education, could be more correct, even in technical matters (Isles and Collado, 1979; Shrestha, 1980). The myth of omniscience needs debunking wherever it is found— whether the belief is held by the educated or by the uneducated. The purpose is to get both groups to accept more collaborative approaches, drawing on what knowledge each can provide the other.

Projects = Development. It has been the practice in development management to draw a sharp distinction between regular continuing programs and development projects, the latter commonly touted as representing the cutting edge of development. Within this framework, development is approached through a series of finite, discontinuous actions with discrete time-bounded outcomes and is dependent on special temporary injections of external funds. Following the Western theory, planning is presumed to be separate from and preparatory to action. The planner does the thinking and draws up the scheme for action. The implementer then follows this blueprint in the best tradition of policy-neutral administration. Institutional structures are regarded as largely fixed and are given little attention; projects preoccupy both planners and implementers If suitable organizations are not available for implementation, the problem is solved by creating temporary project management units.

Projects implemented by temporary organizations to achieve limited, time-bounded results are commonly defended as having positive demonstration effects. The presumption is that the new concepts will be adopted by more permanent organizations once they are proved in the field. In our experience, such "spontaneous" replication by the "permanent" bureaucracy is rare, precisely because, if the concepts represent real innovation, their application will require in-depth changes in the values, structures, and operating systems of the agencies that apply them. The demonstration of a new program idea by itself is unlikely to lead to such changes. These in-depth changes constitute the more difficult part of development innovation yet receive little attention from donors or from recipient governments (Sussman, 1980; D. Korten, 1980).

Within the framework of a strategic management perspective, development projects become laboratories for mutual learning (to improve organizational systems within the bureaucracy and to strengthen problem-solving capabilities within the community) rather than merely circumscribed work units dedicated to producing predetermined outputs and outcomes. This approach poses problems for management, demanding a higher order of commitment and imagination than often found now. We believe that the commonly observed lack of commitment and imagination among the personnel of development agencies is more a consequence of bureaucratic systems which treat creative behavior as dysfunctional than a reflection of any inherent qualities of their personnel.

Sustained and widespread rural progress depends more on moving whole sections of the bureaucracy toward new modes of operation than on the (often temporary) "enclave" progress too often associated with projects. It calls for a strategic style of management which sets directions, creates new visions of the possible, and builds new organizational capabilities to respond to changing conditions. This perspective views development as a continuing process of changing relationships, defines development resources broadly (not just as capital), treats planning and implementation as a continuing, repeated effort to deal with changing obstacles and opportunities, and recognizes the essential role of creative operational personnel working at the point of contact between bureaucracy and the community. This creativity is fueled by close working relationships with intended beneficiaries.

The most difficult part of formulating a strategy for bureaucratic reorientation is devising an effective program of structural change within the bureaucracy, introducing new procedures for project formulation, new criteria for allocating funds or making staff assignments, new personnel systems, more flexible and appropriate budget cycles, new financial and accounting methods, and other policies that support the attitudes and behaviors appropriate to a participatory style of rural development action.

Compliance = Performance. In agricultural bureaucracies, high officials have special problems in monitoring the performance of subordinates scattered about the country. Typically we also find a lack of trust, which is likely to be accentuated by differences in social class, with the top administrators generally coming from higher strata than subordinates in the field. This separation in space and status tends to build distrust, with superiors worrying that their subordinates are lazy and irresponsible. Since the superior cannot be on the spot to observe, he tends to create procedures and reporting forms through which he hopes to be able to keep track of what is going on. Furthermore, each new administrator, reacting to the same difficulties of remote supervision, is likely to want to create his own special forms to provide him with the particular data he considers significant for controlling his subordinates. Rarely does a new administrator eliminate an old form when he introduces a new one. Studies of the work load imposed upon subordinates and of the uses superiors make of the reports are exceedingly rare. We find a natural tendency for paperwork to become so heavy that subordinates may have to spend more time in filling out forms than in getting the job done (Heginbotham, 1975).

Since determining the effectiveness of a professional or technician in the field requires economic and technical data which few bureaucracies can gather or analyze, there is a natural tendency for administrators to use compliance as a proxy for performance. If his reports indicate he is complying with the duties given him by his superiors, the subordinate must be doing a good job. Furthermore, since administrators commonly impose upon their personnel in the field a much heavier load of duties and report-writing than can be fully discharged by a normal human being in an average work week, subordinates learn which reported tasks are given most serious attention by superiors and do their best to simulate compliance on those, while neglecting others. The subordinate's actions reflect what his superior would like to believe instead of his own diagnosis of what the field situation requires.

What we have described is sometimes referred to as a pathology of the bureaucratic organization. "Pathology" is a misleading word here, however, because it generally refers to an abnormal condition. On the contrary, what we have described can be taken as the usual condition that will inevitably prevail unless leaders and organizational planners consciously take steps to meet the characteristic problems found in hierarchies (Downs, 1967).

Premises of Participatory R&D Programs

The premises of participatory R&D organizations contrast sharply with those of conventional R&D organizations. In its most fully developed form, the participatory approach is based on the following set of premises:

(1) Researchers, extensionists, and farmers all have knowledge crucial to the research-extension process.

(2) The best and certainly the most legitimate farm management decision maker is the person who bears the risks of failure and who has the most intimate knowledge of the particular agronomic conditions in question—the individual farmer. The role of the supporting research-extension system is to increase the range of viable options available to the farmer, collaborate in the solution of critical problems, and strengthen the ability of the farmer to make fully informed farm management decisions.

(3) The right choice of technology for small farmers is a function of their personal circumstances and the agronomic characteristics of their land. It may vary from one part of a small plot to another because of differences in microenvironments.

(4) Extensionist and researcher are most effective when they work in such close collaboration with selected farmers that the distinction between their roles becomes blurred: the farmers take initiative in small-scale field trials, assisted by extensionists and researchers, and act as sources of extension assistance for their neighbors; the researcher spends some time working with small farmers to learn from them and to identify with their perspectives. Relationships are horizontal and involvements simultaneous.

(5) The primary responsibility of the agricultural research-extension system is to respond to the needs of farmers for a continuous flow of new technologies by which they can enhance their livelihoods.

(6) Applied agricultural research-extension activities are appropriately organized on a regional or area basis, with an emphasis on developing the full potential of a defined area for sustained agricultural production, to the equitable benefit of its inhabitants. Interdisciplinary teams are the basic unit of organization and polycropping systems provide the focus.

Most experiments with participatory cropping systems have been carried out on a pilot scale, using specially constituted teams working on temporary assignments to devise specific technology packages. To date, little attention has been given to the implications for restructuring a national agricultural research-extension system to make participatory farming systems the standard mode of operation.

Redesigning Agricultural Agencies

The redesign of agricultural agencies along participative lines will require major changes in organization structures and social processes of management. By structures, we mean the formal division of labor and allocation of authority and responsibility to particular positions and position holders. By social processes of management, we mean the pattern of interpersonal interactions and activities through which the work gets done.

It is a common error of organizational planners to assume that, when they have worked out the structure for the organization, their job is done. Structural change is a necessary but not sufficient condition for building a participatory human system. Because in most cases the participatory processes of management cannot be built without a new structural base, let us consider structure first.

Decentralization. A participatory research-extension system must be based upon a decentralized structure. We must distinguish between geographical decentralization and decentralization of power; the two do not always go together. The research or extension organization is always to some extent decentralized geographically in that its activities are never concentrated entirely in the capital city. Nevertheless, traditional administrators often seek to overcome the effects of geographic dispersion through imposing tight control procedures.

The first step toward decentralization would be a study of current reporting forms to reorganize the paperwork so that subordinates have fewer papers to fill out and so that their reports to superiors provide useful information which they can use to guide and evaluate their own work. Such a reasonable reporting system cannot be devised without the active participation of agents in the field, who are, after all, the chief authorities on the nature of their own work.

As experimentation in the regions increasingly emphasizes on-farm work with the active participation of farmers, people in the central office may make important contributions in conceptualizing the new system and orienting those who are to play important roles at the local levels, but any system that depends upon active collaboration among small farmers, paraprofessionals, and professionals cannot be directed by a central authority. Within broad limits, the people who carry out the on-farm experiments must have freedom to make their own decisions.

The new emphasis upon cropping and farming systems also demands decentralization. The head of the maize program, however competent he may be, cannot dictate what variety of maize shall be used in the northwest corner of village X in region Z, because the best-yielding maize variety for the nation, or even for that particular region, may not fit into the local farming system. The timing or labor requirements of the crop may have opportunity costs exceeding the incremental value of the innovation. This problem involves more than monocultural specialization versus a cropping or farming systems strategy. If the central organization has a unit in charge of promoting the development of cropping and farming systems research throughout the country, members of that unit may play valuable roles as advisers to regional personnel, who are developing on-farm farming systems experiments, but the central office people can never have detailed and intimate knowledge of local farming systems and conditions of soil, water, and climate to enable them to make rational decisions regarding local farming system experiments.

Coordination. Coordination requires a combination of structural and process elements. Much of our discussion has involved horizontal relations, that is, relations among individuals at the same organizational level, who need to work together if their tasks are to be satisfactorily carried out, but no individual has formal authority over others. The nature of the problem becomes clear when we consider relations between units of the same organization or between two organizations independent of one another. For example, the directors of research and of extension share a horizontal relationship because generally neither individual has the authority to give orders to the other. How are their activities to be coordinated?

In the traditional agricultural bureaucracy, each administrator has a common superior, generally the minister of agriculture, who is responsible for coordination of the two units. As research directors, supported by professionals in universities and agricultural research institutes, have emphasized the importance of more freedom and protection from the pressures of political patronage, a number of governments have shifted their organizational strategy so as to establish research in a semiautonomous unit. The research director, still reporting to the minister of agriculture, then has more freedom to direct his program than the bureaucrat at his same level in other organizations.

As we have seen in the ICTA case, though the establishment of a semiautonomous research organization favored the development of a strong and innovative research program, at the same time it made more difficult the task of working out collaborative relations with the extension organization, which remains within the traditional bureaucratic structure.

How can more effective interorganization coordination be achieved?

As an Israeli student of rural development has observed, the usual approach is to establish top-level coordination (Weitz, 1971). Government establishes a commission composed of cabinet ministers or their deputies. The commission has the general responsibility for planning and monitoring the supposedly integrated development program. Such a structural element may be useful in achieving some top-level coordination and in providing the rhetoric supporting cooperation among agencies in the field, but the commission generally meets (at irregular intervals) in the capital city, remote from the development area. Even when there is general agreement within the commission on objectives, each agency has its own set of priorities and time and financial constraints, which impose barriers to coordination in the field. When such barriers are encountered, regional administrators can refer the problem to their central office superiors and to the interministerial commission, but that process imposes costly delays and is likely to result in decisions that fail to respond adequately to field problems. Being remote from the local scene in space and social position, commission members are more likely to respond to the politics of the distribution of power allocated to their agencies rather than to the needs of the local situation. In other words, a coordinating body at the top governmental level cannot possibly achieve the sensitive interrelations required at lower levels for an effective common program.

Coordination as a Social Process. Robert Chambers (1974) has argued that coordination and cooperation can be achieved only through developing a management process involving the regional heads of agencies concerned in agricultural and rural development. These key officials meet under the chairmanship of an area coordinator, who reports directly to the provincial administrator. There is an annual meeting for discussion and agreement on plans for each agency for the next twelve months and the budgeting for that period. These meetings enable most of the staff at all levels to have input in the planning process and a broad commitment to its objectives.

The same administrators meet monthly. All who are directly involved in the implementation process present progress reports, identify problems, and discuss possible remedial actions. The coordinator then prepares a monthly management report, summarizing progress and problems encountered and naming those individuals who are responsible for carrying out the actions decided upon. This report is sent quickly to all participants and to agency people at higher levels of government.

Chambers claims that it is customary for people administering development programs to concentrate on plan formulation and budgeting, neglecting the activities that must follow if the plan is to be carried out and the financial resources are to be available when needed. The full sequence of activities needed must include programming (working out detailed assignments of who will do what and when) and implementation (the actual carrying out of the activity). Provision needs to be made for monitoring (establishing an information-gathering system to check progress), evaluation of past performance (to diagnose problems encountered and determine factors affecting success or failure), to be followed finally by reformulation of plans. The last step is of great importance to avoid the common tendency to draw up plans that appear desirable and practical but cannot be implemented. Chambers does not view planning as a once-a-year or once-every-five-years set of tasks but as a continuous process, within which the information brought in from the field through monitoring and evaluation of past performance serves to shape future plans.

According to Chambers, this management system induces regional and area officials to work together across departmental lines. Cooperation among agencies is secured by peer pressure rather than by authority from the top because "no officer wishes to be criticized by his colleagues in a monthly management meeting" (p. 51). The system also appears to have the advantage of making plans and expectations among superiors and subordinates more realistic. Through participation in the annual meeting, the local official has the opportunity to influence the plans that he will have to carry out. He is unlikely to accept without protest unrealistic targets proposed to him by superior officers. If he knows that he will be reporting to his peers each month, he wants to be sure that the commitments he makes at the start will not place him in the position of constantly having to explain failures to meet accepted targets.

In the Special Rural Development Program in Kenya, analyzed by Chambers, the key role is played by the area coordinator, who is responsible for assisting officers at all regional levels in the various ministries or agencies (agricultural research, extension, public works, health, communications, and so on). He serves as a communications link among national, provincial, and district levels and among the various agencies. He is the chief contact with donor representatives, evaluators, and visitors. He draws up timetables and work programs and has the chief responsibility for managing the flow of paperwork.

The coordinator has no authority to give orders to his colleagues in the various departments. His influence depends in part upon his skill in managing the reporting in the discussion meetings, so that he can help people in the field to demonstrate that they are performing effectively.

His ability to exercise such influence depends also upon his support from the chief administrator of the region and the central government. Since the capital city may be far from the development area, it is important that the coordinator have the support of the regional administrator—and that the regional administrator have the ability to provide that support. The viability of the position of coordinator depends upon the degree of decentralization of authority and responsibility to the region. If the chief regional administrator enjoys a considerable degree of autonomy, his support for the coordinator will be more effective than if the top regional administrator is simply passing on orders and instructions from above and reporting to higher authorities.

Organization of Research

There are three main bases governing the structure of the agricultural research organization. The research unit may be organized according to discipline, crops, and geography/area. Cross-cutting these is the dimension of centralization-decentralization, discussed above. Each method of organization solves certain problems of coordination, communication, and so on but also creates new ones.

Specialization by discipline has the advantage of allowing specialists to concentrate on research recognized as high quality in their particular discipline and increases the likelihood that the unit will achieve scholarly recognition. It has the disadvantage of weakening links to production programs or crops, with the result that findings may never be translated into improved agricultural practices.

Organizing according to crops has the advantage of concentrating interdisciplinary attention on the various requirements for genetic improvement and yields of crops identified as national priorities. But it may weaken the unit's strength in specialized disciplines. For example, the plant pathologist who works only on beans may lack the knowledge and skills of one who has in-depth experience with a variety of plants, including beans. Even more limiting, crop-specialized monoculture research rules out research on cropping systems and may leave the program unresponsive to important regional differences that affect the performances of given plant varieties.

Organization by geographical region provides the basis for research on locally specific cropping systems, but it limits possibilities for achieving more basic genetic improvements or for the advances in basic scientific knowledge on which continued technical advances depend.

The structure of an effective agricultural research organization must reflect a balance among all three bases of organizing and will probably involve application of matrix design concepts. To make this concept clear, some background is necessary. One of the first "laws" of organization promulgated by Frederick W. Taylor and his followers in the early twentieth century under the banner of scientific management was the principle of unity of command. According to this principle, every member of the organization below the top must be responsible to one, and only one, boss. In fact, complex organizations are never organized this way, though it took theorists a long time to accept this reality and recognize that it reflected an imperative of complexity rather than a deviation from "correct" principles.

While there may be an infinite variety of patterns within a matrix organizational structure, the essential point is that the matrix design takes into account the necessity of individuals being responsible to and responding to the supervision of two or more individuals. Applying this framework to agriculture, for example, we can have a plant breeder in region X being directly responsible to the regional director of region X but also reporting to a superior in plant breeding at the central office and, if he is working primarily on maize, reporting to a central office superior in the maize program. If this same plant breeder happens to be in charge of an interdisciplinary cropping systems team in the region, he or she may also report to the head of the farming systems unit in the central office.

Similarly, individual specialists might rotate through a variety of assignments, emphasizing different dimensions of their role. For example, the plant breeder might spend time on a regional farming systems team and later be reassigned to a plant breeding program at a national research institution doing basic research within the discipline or in relation to a particular crop.

Possibilities for confusion and conflict in such arrangements are self-evident. But that is the nature of reality in dealing with complex, multifaceted problems—and most problems encountered in rural development are of this variety. It is important to recognize that the messiness inherent in the problem cannot be cleaned up simply by making rules as to who has authority over whom in given situations. Formal rules are an inadequate guide in complex settings, which demand rapid response to new data. Returning again to Chambers (1974), these problems of coordination can be resolved only through the social processes of management. These processes depend on leadership, teamwork, mutual influence, and accommodation, backed up by an organizational culture that supports a consultative style of decision making and makes it possible for people with different interests and responsibilities to arrive at mutually acceptable solutions to problems.

Management of these processes may be facilitated by the formation of overlapping teams, which bring individuals together in different groupings to deal with different dimensions of farming problems at different times. The plant breeder assigned as head of a farming systems team might meet periodically with other farming systems team leaders to share experiences and resolve problems of general concern. He or she might also meet on other occasions as a member of a team made up of the heads of disciplinary and crop groups working within the region to take an overview of regional needs, or of a national team working on the problems of corn production, or of still another team consisting only of plant breeders. It is important that such coordination discussion meetings be held on a regular basis, not only when particular conflicts arise, so that team members can become committed to common objectives and develop a problem-solving technique to arrive at mutually accepted decisions rather than allocating the blame for unsolved problems.

Effective organization of agricultural research requires attention to both structure and process and a recognition of the diverse and sometimes conflicting needs to which the research system must be responsive.

Integrating Research and Extension

One common criticism of the traditional system is that adequate linkages are seldom achieved between research and extension activities. Such systems tend to differentiate sharply among the roles of researchers, extensionists, and farmers according to their status, function, and education and to operate as if the dependence must be sequential and one way: extensionist responding to researcher and farmer to extensionist.

A participatory system needs highly developed expertise as does the traditional system, and yet it calls for a substantially reduced differentiation in status and function as researcher, extensionist, and farmer seek common goals through a team effort in which each is an integral contributor and their interdependence is recognized as reciprocal. It also calls for a substantially revised mix in the relative amount of responsibility given to the professional with the university degree, the graduate of a technical high school or junior college, the paraprofessional, and the farmer leader. The focus in the participatory approach is on substantially strengthening the roles of people in the last three categories. Traditionally, only the professional was thought to be qualified to study a local farming system or to carry out on-farm experiments. Observations in Guatemala and Honduras indicate that many technicians, either in research or extension organizations, are fully capable of gathering and interpreting the information necessary to define and describe a local area farming system and also of carrying out on-farm experiments in active collaboration with farmers. Similarly, many paraprofessionals have a high level of intelligence and competence which, combined with their positions in the local community, enable them to carry out many of the functions performed traditionally by professionals and technicians.

For the new approach we are proposing, advanced technical expertise is only one of the requirements. A narrow but competent specialist may serve a participatory program well as a consultant working only on problems in his or her discipline, but such a person cannot provide the leadership such a program requires. Future leaders will need to build on a strong disciplinary base an ability to think in interdisciplinary terms and to integrate people of different disciplines and social levels into effectively functioning teams.

Structural innovations, such as the placement of research and extension functions under the same leadership and the introduction of formal coordinating mechanisms at various organizational levels, are only partial answers to the integration need. Locating the leaders of the two services, as in Honduras, in offices near each other to facilitate informal interaction can also be helpful but again is only a partial answer. Truly effective integration also calls for attention to the reorientation of training, redefinition of roles, and substantially increased attention to managerial process. There is no simple blueprint for the changes required and only very limited experience to provide a guide. The structures, roles, and relationships appropriate to any given setting can be worked out constructively only within that setting through a long-term participatory process involving the key actors at various levels.

Staffing of Agricultural Agencies

After the designers of the organization of agricultural R&D have conceptualized the activities and the social processes involved in planning, implementing, and evaluating these activities and have designed the organizational structures to support and facilitate them, they need to consider the categories of personnel needed and the division of labor among these categories. We are not assuming that there should be any sharp jurisdictional lines separating the work of one category from another, but it is useful to think of categories as shaped by previous education and experience and by the responsibilities people will carry in the field.

Thinking now only of people who will work outside of the office, it may be useful to use four categories: professional, technician, paraprofessional, and farmer leader. Traditionally, the professional is a person who has a university education and, in some cases, even an advanced graduate degree. The technician (*perito* in some Latin American countries) is generally a graduate of a technical agricultural high school or junior college. The paraprofessional is a local farmer (or farmer's son or daughter) who devotes part or full time to the service of the agricultural R&D program. Generally this person will be young and have more education than average for the community but still not beyond elementary school. The paraprofessional usually has received at least several weeks of education and training in agriculture and rural development (Esman et al., 1980). The farmer-leader is a local man who is an informal collaborator in experiments and development projects. Though not on the payroll, he may receive financial support for attendance at conferences and other meetings at which farmers and government personnel discuss problems.

In the traditional organization, people in top positions tend to evaluate the quality of their organization on the basis of the number of staff people who are university graduates and, increasingly, the number who have graduate degrees. The traditional university education tends to shape graduates who fit into traditional agricultural R&D organizations and are ineffective in the new style of organization. Even if university education were to be reshaped along the lines indicated in Chapter 19, it is important to recognize that the emerging participatory form of agricultural R&D depends much less upon persons in the professional category and much more on the three other categories than does the traditional system.

Leaders of the organization will find that, if they have been selected well, many paraprofessionals have a high level of intelligence and competence which, combined with their positions in the local community, enables them to carry out many of the functions traditionally performed by professionals and technicians.

Linking the Bureaucracy with Farmer and Community Organizations

The emerging new system depends on the participation of small farmers, and, since we cannot expect all farmers to participate equally, it is important for professionals, technicians, and paraprofessionals to work with farmers who are not only open to collaboration with the agency staff but who are accepted and respected by their fellow villagers. In the past, in trying to overcome resistance to change, professionals have sought out those they characterized as progressive farmers, assuming that if they could get a few of these individuals to break out of the traditional mold of their culture, and those individuals got good results, others would follow. All too often, the progressive farmer turned out to be someone above the level of most villagers in education, size of landholdings, and standard of living. He also tended to be a person who was receptive to the ideas and information presented by a prestigious outsider and anxious to please the outsider. By the same token, such a progressive farmer, being already somewhat distant from his fellow villagers, would have little influence upon them.

This new style of organization requires the active participation of small farmers who are informal leaders among their fellows. This is not the place to offer a detailed exposition on how the professional can discover who these informal leaders are. Furthermore, if we did provide such guidelines, we would reinforce the misguided view that the outsider must choose local leaders. The outsider may participate in the selection process by discussing with local people the responsibilities and activities of a farmer leader, but the system calls for local people to play the major role in selecting those whom they wish to lead them in the new approach to agricultural research and development.

Enlarging the responsibilities of nonprofessionals in agricultural R&D does not involve downgrading the responsibilities of the professional. As he delegates tasks to those with less education, he frees himself to observe, advise, and guide technicians, paraprofessionals, and farmer-leaders more effectively.

As we have emphasized, a one-on-one relationship between an agricultural professional and an individual farmer can never be cost-effective. Planners and administrators building participatory programs can expect little payoff from this reorientation unless the bureaucracy becomes more effectively linked with farmer and community organizations.

Where those organizations exist, as in Honduras, the task of achieving effective interorganizational relationships is immensely simplified. Where no such organizations exist, the agricultural agency faces the difficult task of helping the small farmers to organize without making them dependent on government.

Local-level organizations of small farmers—especially when they are linked into area or regional organizations—also generate forces leading toward decentralization of government bureaucracies. If local and regional organizations of small farmers demand action from agricultural researchers and extensionists and bank officials in the field, higher government officials can be responsive to these demands only by allowing considerable freedom of action to regional and area officials. Since small farmers in different parts of the country have different interests and needs, it is impossible for the agriculture-related agencies to serve them effectively when all of the important decisions are made in the capital city. In Chapter 14, we described different types of local-level organizations, which are serving to link small farmers with government organizations in agricultural research and development.

The Learning Process Approach to Bureaucratic Reorientation

The participatory system advocated here cannot be achieved simply through administrative fiat and structural changes. Establishing new roles and relationships can be facilitated by a learning-process approach to bureaucratic reorientation (D. Korten, 1981). This approach is currently being pioneered in two efforts under way in the Philippines—one in the Bureau of Forest Development and the other in National Irrigation Administration (F. Korten, 1981). In each instance, a major national bureaucracy is strengthening its capacity to promote more productive and equitable development based on community-level management of land and water resources—one focused on irrigation, the other on upland agroforestry systems. The processes are evolutionary and agency-specific, and they involve a centrally guided bottom-up process of building new approaches to field operations based on field experience. Although neither deals directly with agricultural research and extension, the methodologies seem well suited to adaptation to those fields.

The learning process is centered within the agency that is trying to achieve its own bureaucratic reorientation by introducing mechanisms that will strengthen its ability to learn from its own experience and to adapt its program and internal structures accordingly. Two primary mechanisms are involved. One is a central working group that manages the learning process. The other is the field-based learning laboratory which is the initial base of innovative activity.

The central working group is chaired by a high-level line official and includes individuals from within the agency and from knowledge-resource institutions, who devote major portions of their time to direct support of the learning process. The group meets at least monthly to review progress, determine needs for special studies, workshops, management systems development, or the initiation of new learning laboratories. Through a combination of foreign donor and agency counterpart funds, it controls resources to contract for research, training, and other inputs crucial to the learning process. These funds also finance supplemental staff not available through regular agency budgets. The working group provides the focus of attention and the resources required to facilitate a change process in a large organization.

One or more learning laboratories provide the sites at which specially chosen agency personnel are encouraged to work flexibly with local citizens to evolve new relationships and action programs that build community capacity to manage local resources in response to the needs of community members.

Activities carried out in these learning laboratories are documented by trained observers who attempt to view the experience from both agency and beneficiary perspectives. These "process documentors" provide monthly reports to line operating personnel and to members of the working group to facilitate their reflection and action on the problems encountered and the successes achieved.

The concerns of the two groups of users are different. Those administering the agency focus on the resolution of individual problems. Working group members are more concerned with using the data to determine what new competencies must be developed and institutionalized within the agency to respond effectively to the problems commonly encountered in the field.

Within the learning laboratory, plans and work methods are subject to revision at any time as the team learns more about needs and opportunities from operational experience. Improvements worked out in the field by this team are examined by the central working group to assess their broader implications for agency policy and structure. The learning process evolves through three stages—each with its distinctive learning agenda (D. Korten, 1980).

Stage 1: Learning to Be Effective. In Stage 1 the agenda focuses on what is required for building community capacity to manage a particular resource equitably and productively for the benefit of its members. Attention centers on the community and its needs. Work in the learning laboratory is likely to be staff-intensive during this stage, long lead times may be involved, and many errors in initial assumptions will be revealed. Success is not measured by the absence of error but by constructive responses to errors. For a learning process in agricultural research-extension, this stage might concentrate on learning how an interdisciplinary farming systems team can work collaboratively with a community to identify needs and opportunities for farming systems improvements relevant to different categories of farmers, support de-

velopment of a community-based research capability, and use local
social mechanisms to disseminate findings. Some of the agency's most
highly qualified professionals might be involved in learning laboratory
field operations during this state. Their involvement would be expensive relative to the benefits received by project area beneficiaries and
would not be directly replicable on any significant scale. But, properly
monitored and analyzed, the learning of Stage 1 provides a necessary
basis for moving to State 2.

Stage 2: Learning to Be Efficient. In State 2 the learning agenda
centers on developing simplified problem-solving routines suitable for
large-scale application by agency personnel of average competence.
These are then tested and further refined in other learning laboratories.
Attention centers on operating-level procedures. During this stage,
leaders give special attention to potential conflicts between the new
methodologies and existing agency structures, procedures, and staffing
patterns. Growing numbers of agency personnel are involved as new
learning laboratories are established to develop the pool of human
resources skilled in the new methodology needed to perform training
and supervisory tasks in Stage 3.

Stage 3: Learning to Expand. One of the most difficult tasks in
Stage 3 is to restrain the impulse to move immediately to national
replication as soon as workable methods are developed in Stage 2. If
the new methods and styles of operation developed in Stages 1 and 2
are indeed a substantial departure from normal organizational routines,
they will depend on the development of new skills, acceptance of new
roles, and reorientation of management systems. Such changes cannot
be achieved overnight. Attention now centers on the structure, systems, and institutional culture of the operating agency. What changes
are essential to support the new mode of field operations? How can
these changes be achieved? The central working group learns to deal
with these issues, through new learning laboratories focusing on major
problems of organization structure and management processes.

In the Philippines irrigation case, regional offices have the responsibility first to establish one locally managed learning laboratory per
region to build expertise needed to introduce the methodology gradually into each province within that region. Concurrently, the central
working group monitors progress to act on needs for further changes in
supporting structures, management systems, procedures, training, and
so on.

In an agricultural research-extension system, Stage 3 learning would
address the need to achieve decentralization, building incentives to
encourage collaborative work with communities, installing lateral inte-

grating mechanisms, and otherwise sorting out relationships between agricultural research and extension agencies. In a large, complex system this process might take ten years or more. Not every agricultural research-extension system has leadership prepared to make the necessary commitment to such a process, but our experience suggests that such leadership is not so rare as might at first be assumed. Work in systems that have such leadership can provide both the inspiration and the models for others to follow.

Part V

Implications for Research, Education, and Government Policy

We have sought to demonstrate that involving small farmers in the advances of research and development cannot be accomplished simply by tinkering with conventional organizations and programs. This book makes the case for a fundamental reorientation of organizational strategies, policies, and programs.

Recognition of the magnitude of changes needed may lead us to pessimism regarding prospects for any major improvements. On the other hand, we are not simply imagining the rural world as we would like it to be. In Latin America, Africa, and Asia, we have found creative and innovative farmers and professionals working together in establishing pilot models of parts of future participatory systems of agricultural research and development. Furthermore, within conventional organizations and programs we find important elements of strength, which, when integrated into more participatory systems, can accelerate progress in agricultural R&D.

So far, our aim has been to examine elements of the emerging participatory system now in action in various developing countries and to visualize the nature of such a participatory system if all the parts were put together. In Part V we turn to a discussion of the direction of change in research, education, and government policy required to build the participatory R&D system of the future.

In Chapter 17, "Implications for Plant and Animal Research," we stress the need to provide agricultural professionals with more interdisciplinary and field experience with small farmers.

For decades social scientists have complained that their potential contributions to agricultural research and development have not been recognized by those in charge of R&D programs. On the other hand, we find encouraging signs of progress in integrating social scientists into such programs. Furthermore, we argue that the small role played by social scientists in the past resulted in part from deficiencies in their own theories and methods. Since we now see these deficiencies being overcome, we forsee a stronger integration of the social sciences in

249

future agricultural R&D programs. Chapter 18, "Reorienting the Social Sciences," considers this changing pattern.

Innovative agricultural scientists often complain that, when a graduate of an agricultural university comes to them, they have to begin by making the newcomer unlearn much of what he or she has learned during formal education. If this costly unlearning process is to be avoided, there is a need for a major reorientation in educational programs—a subject discussed in Chapter 19, "Implications for Education."

Governments have large and pervasive influences across a wide range of agricultural organizations and activities. Chapter 20, "Implications for Government Policy," discusses what governments can do to create and support the participatory agricultural R&D systems of the future.

17

Implications for Plant
and Animal Research

We have seen in previous chapters that small farmers constitute half or more of the rural population in most LDCs and that they have to operate under economic constraints caused not only by the small size of their operations but also by limited credit, limited availability of commercial inputs, and disadvantages in marketing. These constraints tend to diminish their ability to undertake risks in their choice of farming enterprises and their purchases of inputs designed to provide maximum production. They also cause small farmers to depend as heavily on family labor as possible and to place a high priority on production of food for family subsistence. We thus have to adapt the methodologies of agricultural research that have been found effective in general to the special needs of smaller farmers, recognizing existing institutional limitations and activities that affect land tenure, availability of credit, inputs, and marketing opportunities, including prices of farm products.

We have also seen that the small farmer's economic opportunities depend as much or more on his natural environment—his climatic and soil resources—as on his socioeconomic environment. The amount and pattern of rainfall distribution, the temperature and light regimes, the physical and chemical properties of the soil, the terrain that he works all determine his practical choices for cropping patterns and animal enterprises. Thus both the limitations and the opportunities for economic improvement are specific to the site that the small farmer occupies. In established small farmer areas, the farmers have adjusted their land use over a long time to the limitations and opportunities encountered in the sites. In new colonizations there may not be an established pattern. In the former case, research seeking to improve the productivity of the area should depend heavily on previous local experience. In the latter case, lacking that experience, the initial research may have to depend more heavily on experience in analogous areas.

The following discussion focuses on four aspects of research and development programs concerned with plants and animals in developing countries: national policies and programs, site-specific farming systems, transfer of research results to analogous areas, and inter-disciplinary research and development activities.

National Policies and Programs

Most, if not all, developing countries now have planning offices at the top level of the political structure with responsibility to provide continuing plans for national development of industry and agriculture. These offices coordinate, usually by developing three-, five-, or ten-year plans, the projections made for attainment of economic and social goals by the various government departments and ministries, taking into account available internal and external resources and opportunities. The agricultural component of the national development plan is usually prepared within the ministry of national resources or of agriculture and animal husbandry, but it is limited by the national priorities and policies for industrial and urban development, rural development, agrarian reform, and expansion of agricultural areas as well as by foreign exchange and price and market policies.

National priorities for agricultural research programs frequently stress the following approaches: research to increase production of specific food crops and animal products in order to satisfy the needs of a growing urban population and to decrease foreign exchange problems; adaptation of technologies for crop and animal products that have been found successful in industrialized countries with highly developed infrastructure; and expansion of agricultural area and irrigation rather than intensification of land use through improved management systems. The technical assistance from international agencies and from bilateral development programs of donor nations has largely been oriented in support of these emphases. Recent recognition of the important role of small farmers in production of basic foods and the importance of improving the well-being of the rural sector in developing countries has usually not been accompanied by changes in planning for agricultural programs that recognize the special needs of this group so that it may participate more fully in the attainment of the national goals for agriculture.

The problem is to provide mechanisms whereby the objectives of the national development plans that require the participation of small farmers in production activities can fit into the farming systems of small farmers in specific regions. The first step in solving this problem is acquisition of knowledge of the farming systems in the different agricultural areas of the country. The second step is determination of special incentives and supports that are needed to obtain the participation of the farmers in the regions best adapted to the planned production programs such as credit, price support, and marketing provisions. In taking both of these initial steps, the planning body needs the participation of local leaders in the farming areas under study as members of the group making the studies and evaluations.

Site-Specific Farming Systems Programs

Previous chapters have dealt in detail with the socioeconomic and ecological constraints that result in a need to develop farming systems programs pertinent to the small farmers in specific locations. Farming systems research provides a holistic approach to the problem of fitting animals and crops to the environment of the small farmers, making improved technology an option—a variable—rather than considering it as the essential path to the attainment of the farmer's goals. Figure 17.1 provides a schematic representation of determinants of the farming system (Norman, 1980). It deals with the socioeconomic constraints under the heading *Human,* and separates the infrastructural and institutional factors under the heading *Exogenous* from the farm family constraints under the heading *Endogenous.* The ecological constraints are dealt with under the heading *Technical.* Research in developing countries, as it has emerged during the past decade in Africa, Asia, and Latin America,

> recognizes and focuses on the interdependencies and interrelationships between the technical and human elements in the farming system. . . .
> The primary aim of the FSR (farming systems research) approach is to increase the overall efficiency of the farming system; this can be interpreted as developing technology that increases productivity in a way that is useful and acceptable to the farming family, given its goal(s), resources and constraints.   [Norman, 1980:5]

As Norman has stated, farming systems research is not a substitute for conventional commodity-oriented agricultural research but is complementary to it.

Norman has found that the successful farming systems research undertaken in recent years involves four stages: description or diagnosis of present farming systems, design of improved systems, testing of improved systems, and extension of improved systems. The first three of these four stages are encompassed in the ten steps described below for the Central American Small Farmer Cropping Systems Program of CATIE (1979). These procedural steps, in turn, are similar to those that are taken in the Asian rice-based farming systems programs of IRRI

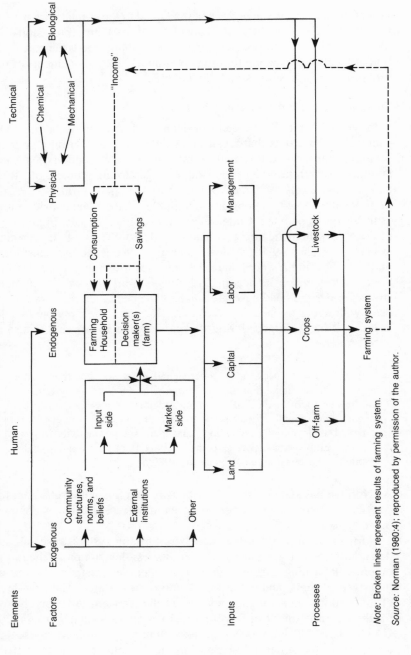

Figure 17.1. Schematic representation of some determinants of the farming system

Note: Broken lines represent results of farming system.

Source: Norman (1980:4); reproduced by permission of the author.

and those incorporated in the Caqueza project in Colombia (Harwood, 1979; Zandstra, Swanberg, Zulberti, and Nestel, 1979). Thus there appears to be remarkable agreement among those who have taken the leadership in farming system work in tropical Africa, Asia, and Latin America, and their experience should be useful to directors of research embarking on new programs.

Step 1: Identification of Goals and Purposes. The goals of this research need to conform to the long-range plans of the governments involved for development of their agricultural sectors. Often three general goals may be involved: improvement of foreign exchange, improvement of national food supply and nutrition, and improvement of the income and welfare of the rural sector, particularly the rural poor. The priorities in dealing with these goals should be clearly identified. The relative importance of these goals determines, in a preliminary way, the priority crop and animal production programs, the agricultural zones to receive priority attention, and the combinations of infrastructural and agricultural technology approaches to receive attention.

Step 2: Selection of Areas for Study. Within the agricultural zones to receive priority attention, specific areas (sites) in which studies of farming systems are to be concentrated are selected. They should be representative of the rural communities, their accessibility to input, markets, and technical assistance, and their climatic and physical environment.

Step 3: Inventory of Documentary Information. The third step is a careful inventory of the information available on the local socioeconomic and physical environment in the study areas. This material is obtained from sources such as censuses, crop and animal production data, and price information, as well as documents on climate, weather, soils, and ecology.

Frequently the inventory reveals a large amount of pertinent data in many of the subject categories at a national or regional level. In some of them, data may be spotty, of short time span, or absent. Sometimes it is possible to fill in these gaps by obtaining cooperation from national or international agencies.

Step 4: Local Surveys and Case Studies. Complementing the inventory, a local survey in each study area should be made with the participation of representatives of the national agricultural research and extension agencies and local people. Because of the limitations of time and money, it is usually not feasible to use the most rigorous random-sampling procedures in this survey. With the help of national extension and research personnel, it may be possible to select rural communities that are representative of the study area and to select farmers within each community whose operations are typical. They should cover the range in size of operation and dominant farming systems. Selected farmers are interviewed about their farming systems, production, and costs. One or more of these farmers in each area should be selected to provide weekly information on operations over an entire year (case studies).

Step 5: Conceptualization and Planning of the Research to Be Done. Making use of the data obtained from the inventory and the surveys, conceptual models describing the principal farming systems and the major problems encountered by the farmers should be developed. The problems should be classified according to those requiring external institutional help, those that can probably be solved or alleviated by changes in farm management or inputs, and those that appear not to have practical solutions. Concentration of the research should be on problems with apparent local solutions. This work should be done cooperatively by research and extension personnel and should be reviewed in detail by representatives of the farmers who provide the survey information.

The purpose of this step is to provide a logical framework for the orderly collection and analysis of data useful to farmers who must choose among alternative practices for their farming systems. The framework must depend upon reliable data, and in turn the conceptual frame should reflect the realities of local farming systems.

Step 6: Farmer Trials. Farmer trials are simple tests of modifications of normal practices within farming systems for the purpose of overcoming limiting factors that affect productivity. Carried out on small farms with the active participation of the farmers, their success or failure is measured by comparing the production and its cost under the trial procedures with that under the normal procedures used by the farmer. Few comparisons are made on a given farm, and replication is minimal (usually not more than two or three replicates on a farm). Additional replications of a particular comparison should be made on several farms of the area on similar soils and with similar farm management practices in order to measure the effects of location and other uncontrolled variables.

Each farmer trial may involve simultaneous comparisons of several changes whose combined effects can maximize increases in productivity and net income (for instance, the use of an "improved" variety together with increased fertilization and a higher level of insect and desease control). Thus it becomes difficult, if not impossible, to measure interactions of variables in the farmer trials. In all farmer trials the effect of the package of change is evaluated in terms of net income and labor demand as well as of production.

Step 7: Component Experiments and Studies. At the same time that farmer trials are conducted, somewhat more sophisticated experiments with varieties, spacing, fertilizer levels, pest control, and other factors are carried on in the area in fields under the control of the research worker. The purpose of those experiments is mainly to screen materials and treatments for their usefulness in the area. Those that show promise are then used in farmer trials. It is appropriate to design these studies so that the more significant interactions may be evaluated. Selection of the materials and treatments to be studied should be rigorously limited to those that appear to have application in solving important local problems of production and net income. For this purpose, the criteria for selection need to be clearly defined and strictly adhered to.

Finally, within the category of component studies, highly complex experiments designed to elucidate basic principles of farming systems that are useful in the area are appropriate only at a regional or central experiment station. Even there, the questions dealt with should be addressed as simply and directly as possible, and the layout of the experimental plots should be flexible, so as to permit changes from year to year.

Step 8: Analysis and Interpretation of Results. The results of the farmer trials and the complementary experiments should be summarized by those in charge of the individual area research and analyzed and interpreted by the central research group with a minimum of delay, so that tentative conclusions may be drawn about the most promising modifications of the prevalent farming systems and about the directions for the research for the coming year. To make this procedure possible, the plot layouts and data collection procedures for all of the field research should be as uniform as possible.

Step 9: Applications of Results to Planning Future Work. The research results and interpretations for each year should be discussed with the cooperating national technologists and farmers. Along with professionals, farmers should participate in the planning of the coming year's research work.

Step 10: Evaluation of Research Results. Each year the research results should be evaluated to determine their fulfillment of the goals and purposes of the program. Since farmer adoption of the new technologies is the ultimate test of the on-farm research, farmers must participate in the evaluation process.

Transfer of Research Results to Analogous Areas

Since the conduct of site-specific farming systems programs, as described above, is time-consuming and expensive, it is important that the research results should be transferable to untested analogous climatic and socioeconomic situations as a basis for planning programs in those areas. For transfer to be possible there must be accepted criteria for grouping farmers into types of farming areas, for effective low-cost techniques for appraising existing farming systems, and procedures for data-taking and analysis that permit comparative evaluations.

Collinson (1972) has chosen three criteria for grouping small farms in Africa into farming areas: a pattern of climate and soil over which production opportunities and improvement possibilities are the same; a common tribal background giving homogeneity in motivational patterns, social tradition, and agricultural practices; and limited variation in the man/land ratio. He suggests that climate and soil data can be tied to existing cropping so as to give an indication of likely changes in productivity of crops grown in the area and identification of the ecological zones under consideration. The FAO agroecological area studies of major crops (FAO, 1978) provide global data on the subject, which complement the available local census and other secondary information available. Knowledge of common cultural backgrounds likewise is available from secondary sources. Surveys may be necessary to obtain information on man/land-water conditions.

The problem of effective low-cost techniques for appraising existing farming systems may be more difficult, primarily because of the tendency in each of the cooperating disciplines (such as agronomy, sociology, and economics) to wish to evaluate rigorously all of the available information pertinent to the field sites. The result may be the amassing and analysis of much more data than necessary for the appraisal required. As Chambers has noted, "What is needed is to develop methods that are quick and cost-effective in terms of the trade-offs among quantity, accuracy, relevance, timeliness, actual use of information, and the costs of obtaining it" (Chambers, 1980:3). Hildebrand, in his initial appraisals in work with ICTA, Guatemala, depended on what he referred to as *sondeos,* or rapid local surveys made by an agronomist, a farm management economist, and a sociologist (Hildebrand, 1977). This abbreviated use of the more rigorous conventional techniques has gained acceptance. Chambers (1980:1) warns, however, correctly, against the pitfalls of "rural development tourism"—brief visits by urban-based professionals—as the basis for appraisal. The CIMMYT manual, *Planning Technology Appropriate to Farmers: Concepts and*

Procedures (1980), provides an excellent framework for the description of farmers' circumstances and a basis for identification of research priorities.

The use of standardized procedures for data-taking and analysis is an obvious way to provide the most efficient comparison of information within and between site-specific farming systems projects. It is also important in providing for the use of such information as a basis for new studies in analogous areas. As this information accumulates from programs in different ecological and socioeconomic areas, uniform reporting becomes increasingly important.

The transfer of research results to analogous untested areas does not eliminate the need for site-specific research in those areas. Rather, the recognition of the similarities (and differences) between tested and untested areas provides guidelines for focusing attention more rapidly on improvements in management and technology that are acceptable and promising in this local situation, thereby saving time and expense. Use of linear programming to extrapolate findings in one site to other sites cannot be depended upon to yield reliable predictions, in view of the large number of variables and interactions that actually exist. On the other hand, within specific sites, the use of simulation approaches may well be justified to explore the probable effects of interactions that have been found to be of dominant importance.

Interdisciplinary Research and Development Activities

Throughout this book we have stressed the interdisciplinary nature of farming systems research and the need for its recognition in the planning, execution, evaluation, and extension activities concerned with it. The disciplinary fields most involved are agricultural economics, agronomy, plant breeding, soils, agricultural engineering, agroclimatology, animal husbandry, rural sociology, social anthropology, and government. Given this array of disciplines, it is essential that there be a clear statement of goals and objectives for each interdisciplinary program and a procedural attack that identifies the roles and responsibilities of the participants representing the different disciplines. Participants should have a broad appreciation of the complementarity of the fields and the need for a team approach to the programs in which they are engaged. Since each program is unique and the contributions expected of the different fields differ in the planning, execution, evaluation, and extension phases, the appropriate leadership may be different in different programs and different phases of the same program. That agricultural development deals with a continuum from subsistence farming to commercial farming, with intermediate stages of mixed subsistence-commercial farming, is stated by Collinson (1972) and elaborated by Harwood (1979:20). In providing baseline data for initial planning of farming systems programs, Greenwood (1978) has stressed the need to understand linkages among communities, regions, and national institutional structures and the reciprocal interactions at these three levels. Government and sociology participants who can assess these matters are necessary at this stage as well as in the evaluation. A farm management participant is necessary not only at these stages but also in the execution and extension stages. An agronomist with considerable knowledge of plant breeding, soils, and agroecology likewise is necessary at all four stages. The contribution of participants specializing in animal husbandry and engineering will depend upon the nature of the program. Where animals are an important part of the existing or potential farming system a full-time participant is essential; otherwise, availability as a consultant may suffice. Where irrigation or farm machinery are important in existing or potential farming systems, an agricultural engineer may be needed full time; otherwise as a consultant.

Organization of National Farming Systems Programs

Although it may be possible to assemble talented expatriate teams for programs with special financing by international or bilateral foreign agencies, and in this way to initiate strong farming systems programs as short-term enclave operations, it is essential that from the beginning they contain counterpart participants from the national institutions designated to carry the programs over the long run. Three difficulties are most commonly encountered: lack of personnel with the necessary motivation and training; organizational separations of the personnel needed for the program; and inadequate financial and technical support for the work of the designated national participants. An overall necessity is a strong national commitment at the top political level to the goals and objectives of the program and a stable government that can foster its continuation through the inevitable changes of personnel within the government. Since the interdisciplinary approach cuts across departmental and agency structures and complements the conventional specialist emphasis of most agricultural research (Dillon, 1976), national farming systems programs should be led by highly respected individuals and supported by a separate budget that provides for the conduct of the interdisciplinary activities and for the training of technologists and the participation of cooperating farmers.

The recent commitments to farming systems research cannot be expected to provide sudden dramatic results comparable to the highly publicized ones of the Green Revolution. The creation of interdisciplinary teams of technicians to work with small farmers in planning and carrying out the programs takes time. Perhaps a decade is the minimum period necessary to make this approach viable within the context of the present organization of agricultural research in developing countries.

Although it may be tempting to suggest radical reorganization of the structure of the agricultural research institutions of developing countries in order to give priority to farming systems programs adapted to the needs of small farmers, an evolutionary approach appears more likely to succeed in the long run. Given an appreciation of the importance of such programs, the most important initial step is to provide for them within the present structures, to complement the conventional reductionist approaches, which concentrate on controlled experiments using disciplinary units to isolate and solve single problems. Where trained staff and budget are limited, the deployment of technical specialists from disciplinary departments to a farming systems program is more economical than employing a full complement of specialists for each program. In addition, the advantage to the individual specialist of retaining his professional identity in a recognized discipline makes it easier for him to interact with his peer group in publications and exchanges of information. Finally, some of the specific problems dealt with by interdisciplinary programs are best solved by conventional experimental approaches. By setting up the farming systems program with complementary leadership and budget, it is to be expected that a productive balance between the disciplinary and interdisciplinary programs of research will evolve and that their association will result in an enrichment of the total research output, both in quantity and quality.

A further argument for an evolutionary approach to the adjustments within existing research agencies, as contrasted with drastic reorganization to give dominance to farming systems programs, is that regional interdisciplinary work brings together biologists, physical scientists, and social scientists, whose technical languages and methodologies may be very different. Bureaucratic separations of the departments within ministries complicates this process; such departments have to be bridged administratively if the necessary coordination is to be accomplished. Since details of organizational structure of the ministries are unique to each country, the ways in which the national interdisciplinary farming systems program can best be brought to life and nurtured may differ markedly from country to country.

External Budgetary Support and Expatriate Personnel

Developing countries depend heavily on external budgetary support and on personnel provided by international agencies or bilateral contract programs with developed countries for the initiation of farming systems programs for their small farmers. Two common difficulties resulting from this dependence are that the program support is provided for fixed periods, usually three to five years, with opportunity for renewal for a second similar fixed period and that the expatriate personnel frequently are products of agricultural institutions organized by disciplinary departments, largely carrying on conventional experiment station research and training oriented to the needs of industrialized economies.

The fixed-term limitations place stress on the new farming systems program to produce visible results within a very short period of time. Since it takes a year or more to provide a new program with staff, facilities, and materials and at least another year to provide the necessary in-service training, contacts, and reconnaissance in the field, a three-year fixed term under the best of circumstances provides no more than a crop year of critical field work with farmers. This period is obviously too short to expect meaningful research results.

Conclusions

National policies and programs for agricultural research generally emphasize increases in production of specific crops and animal products, technologies that have been found to increase production of these crops in industrialized countries, and expansion of the total area and increase of irrigated area devoted to these crops and animals, rather than improved management systems. These emphases frequently do not recognize the special needs of small farmers or provide the incentives necessary to get them to participate in the development programs. The policy and program problem, therefore, is to provide mechanisms whereby the national plans fit into the farming system of small farmers in the regions to be affected.

The development of improved farming systems for small farmers depends upon a holistic approach to the problems of fitting the farm enterprises into the total environment of the farmer. Successful research programs undertaken in tropical Asia, Africa, and Latin America depend upon active participation of the small farmers in the planning and execution of the work.

The transfer of results of successful programs to analogous areas can be helpful as a basis for planning only if there are accepted criteria for grouping farmers into types of farming areas, effective low-cost techniques for appraising existing farming systems, and procedures for data-taking and analysis that permit comparative evaluations. Some progress has been made in these directions, but there is a good deal more to be accomplished.

There is now general recognition of the interdisciplinary nature of farming systems research focused on small farmers and the need for its consideration in planning, execution, evaluation, and extension activities. Within the broad disciplinary areas of the natural sciences, ecological and agronomic approaches are tending in the direction of agroecology. Within the social sciences, anthropology and sociology are becoming better integrated with macro- and microeconomics. The gulf between natural and social sciences is narrowing as a result of better understanding of their complementarities. Within the context of specific programs for improvement of small farmer farming systems, the team approaches needed will differ with the limiting ecological and social constraints encountered, and the composition of the interdisciplinary teams should recognize these differences. There is need for further progress in these directions.

National farming systems programs can be successful in the long run only if they are based on long-time commitments by the governments initiating them. Expatriate participation based on external funding may be necessary and desirable in the early stages, but from the outset it should serve to mount self-sustaining national activities, staffed and funded within the framework of national institutions. The institutional framework within countries differs, and the new emphasis on farming systems approaches should evolve within the framework, to the extent possible, rather than competing with it. Such a process requires time, and a decade or more may be needed for it to evolve.

18

Reorienting the Social Sciences

The new approach to agricultural research and development advocated here will require interdisciplinary collaboration between and within the agricultural social sciences and communication between scientists and farmers in the design and conduct of research. Some argue that this is not a new but an old philosophy. Thus a first step in this chapter will be to review previous interdisciplinary efforts. Then we will examine the disciplinary focus of the social sciences in development-related research in the 1950s and 1960s. The controversy over the benefits of the Green Revolution that erupted in the late 1960s and early 1970s led to assessment of the constraints to agricultural development and the role of social sciences in alleviating these constraints. After discussing the reorientation of social science research, we conclude the chapter with an agenda for the social sciences—a set of problems that will require both disciplinary and interdisciplinary research of the highest caliber.

Interdisciplinary Research in the United States

Early in this century there was little if any formal interdisciplinary research in the United States. The research staffs of the colleges of agriculture were small. Thus there was a good deal of informal interaction among disciplines, just as there is today in the international agricultural research centers. In such an environment each individual's disciplinary research is likely to reflect some influence of other disciplines. Furthermore, because the rural social sciences were developed in the agricultural colleges, many of the early practitioners came from a farm background. Communication between farmer and scientist did not present the obstacle that it does in the developing countries today, where there is at least a perceived intellectual and social gap between farmers and the researchers. The very use of the term "peasant" tends to emphasize this class distinction.

Perhaps the earliest formal interdisciplinary research involved plant pathologists and plant breeders working on cereals and potatoes. This work dates back to the 1910s and 1920s, when the only way to deal with blight and rust was to work together. Plant breeders such as H. H. Love at Cornell used a team approach in Asia in the 1920's, and the Rockefeller program initiated in Mexico in the 1940's, which was the interdisciplinary precursor of CIMMYT, emphasized plant pathology together with breeding.

In 1925, John D. Black, a well-known Harvard economist, speaking at a meeting of the Association of Land Grant Colleges and State Universities, emphasized the role of economics in agricultural research: "If I had time to do it, I could show that most of the problems of an economic nature which confront our [experiment] stations require a combination of economic and natural science analysis. All problems of the relation of input to output are of this nature, all fertilizer input studies, many feeding experiments, and cultivation experiments" (Black, 1959:622).

Among the earliest examples of interdisciplinary research involving animal scientists and economists is that of Einar Jensen et al. (1942) on input-output relationships in milk production. The two disciplines worked together on design and analysis of feeding experiments. Like the early work on pathology and breeding, this project was integrated from the beginning. In the 1950s and 1960s, a great deal of interdisciplinary research was conducted between agricultural economists and biological scientists, particularly at Iowa State and Michigan State universities (see, for example, Hoffnar and Johnson, 1966). The degree of formal collaboration among researchers varied considerably from project to project. In some cases, individuals worked together under a broad mandate that allowed each to pursue his or her own interests. In other cases, a tightly managed group worked under a project director, meeting frequently to discuss the design and execution of the research and integrating the results into a coordinated product. One of the most ambitious of these efforts involved fourteen Cornell faculty and graduate students from seven disciplines in a study of nitrogen and phosphorus balance in the environment (Porter, 1975). It is one of the very few projects in the United States in which agricultural economics and rural sociology were both involved.

Joint research between agricultural economics and natural sciences and engineering is well documented in the surveys of agricultural economic literature (see Martin, 1977). Relating agricultural economics in a systematic way to the other social sciences does not have a parallel record. Earl R. Swanson (1979: p. 849) notes that working with other disciplines, usually in the social sciences, in a problem-solving mode has been a recurring theme for at least two decades in the presidential addresses, invited papers, and fellows' lectures at the annual meetings of the American Agricultural Economics Association.

Despite these admonitions and modest efforts in interdisciplinary research, as the disciplines have grown, communication among them appears to have declined. Lack of communication is often more marked in the liberal arts than in the agricultural faculties, which have a stronger tradition of applied and problem-oriented research. Those inter-departmental projects that do exist are commonly organized in such a way as to enable each individual scientist to work independently, there-by reducing the time costs but also the benefits of doing truly inter-disciplinary research. The professional gains from specialization are too high to be sacrificed casually, and the professional rewards from interdisciplinary work have generally been inadequate to attract many serious scholars.

Disciplinary Orientations to Research on Agricultural Development in the 1950s and 1960s

The work of the agricultural economists and of the other social scientists has taken very different paths. We will discuss the latter group first, listing among "other social scientists" sociologists, anthropologists, social psychologists, and political scientists with perhaps an emphasis on the first two. We recognize that such a grouping does injustice to the distinctly different attributes and orientations of each discipline, but these two disciplines seem to have had a predominant interest in the psychological orientations of peasants and in the social processes required to change these orientations in the direction of so-called modernization.

The prevailing views of anthropologists and sociologists a generation ago were summed up in an influential book reporting on a conference on rural development at MIT:

> The behavorial scientists had their various special diagnoses, but were, after the habit of their kind, less positive as to solutions. They accused the first two groups (agronomists and economists) of neglecting the special values held by the traditional peasant and overrating the importance of technical knowledge and economic incentives. The rural villager, they said, is a prisoner of his culture and his history, suspicious of change or innovation, not accustomed to taking the risks involved in producing for a market, and is, therefore, differently motivated from the commercial farmer. They tended to take the gloomy view that there was not much hope until the whole structure of rural society was rather radically altered

and its values changed through fundamental education, a breakup of the extended family, and the spread of mass communications. [Millikan and Hapgood, 1966:p.vi]

Paradoxically, although highly critical of what they saw as the prejudices of natural scientists against social science, behavorial scientists in their studies of resistance to change were implicitly assuming that innovations recommended to small farmers by agricultural professionals were always right. If the farmers rejected what was good for them, this passive resistance must be grounded in social and cultural influences that bound them to traditional ways. Thus arose what we have called the myth of the passive peasant.

Many social anthropologists in this era were ambivalent about modernization per se. Impressed by the complex interdependencies among elements of the indigenous cultures, they were worried over the potential destructive effects of introduction of change from outside. They, too, thought in terms of resistance to change but considered it a healthy phenomenon in many cases.

Those anthropologists who accepted the inevitability of externally introduced changes sought for ways to guide and shape these changes so as to minimize their destructive effects and to maximize their potential benefits. This line of work offered more promise of practical application, and yet the style of research tended to limit its usefulness.

Characteristically, the social anthropologist thinks of spending at least a year in any field study of a community so that he can immerse himself thoroughly in its culture and social organization. Although some plant and animal scientists are willing to recognize the value of the insights of a social anthropologist regarding "his village," they naturally find it difficult to see how such a time-consuming methodology can contribute to practical results beyond any single village.

Diffusion Research. The misdiagnosis implied in the myth of the passive peasant led to very active research on the diffusion of innovations, a scientific dead-end street well described in the following statement by two of our colleagues:

Communication became the key variable in diffusion analysis, and the study of communication processes was deemed necessary for understanding agricultural modernization (Rogers, 1969). From this perspective, development was most often viewed as the sum of many individually made decisions concerning the acceptance or rejection of innovations. Lack of information generally was assumed to be one of the major factors limiting modernization, and consequently, attempts were made to identify communication barriers which restricted the innovation diffusion process. Research questions accordingly tended to address the nature and characteristics of the innovation; the perceptions, values and motivations which comprise the individual's decision-making framework; and the nature and characteristics of the adopter himself. Likewise, communication channels and local opinion leadership were identified in the effort to discover the most efficient means of reaching the target population. Recommendations ensuing from this approach generally concerned the most effective channels of communication, the types of persons most likely to be receptive, and construction of appropriate messages for compatibility with local culture, values and aspirations. In most cases, these recommendations led to efforts to improve communication, and were frequently operationalized as increases in the quantity and quality of agricultural extension agents (Fliegel, Roy, Sen, and Kivlin, 1968; Leagans, 1971).

The assumptions made by researchers employing the diffusion perspective suffered inherent limitations. First, technology was assumed to be available and relevant. Practices which resulted in increased productivity under experimental conditions at a research station were frequently assumed to be applicable throughout the existing agricultural system (even though physical conditions usually vary considerably). Likewise, the existence of the infrastructure necessary to support the innovation, e.g., input, markets, credit, transportation and storage was likely to be taken for granted. Second, the suitability of a particular technology for various types of farms or groups of farmers was normally not questioned, thereby ignoring the possibility that within certain social settings technology may enhance rural welfare, while in others its impact may be detrimental.

Third, the diffusionist focus on the role of individuals in the communication process usually placed such considerations as social structure and institutional influences in a secondary plane. The individual was seldom seen as belonging to a system, be it agricultural or social. Individual characteristics rarely came to be viewed as group patterns or structural features. Innovators may have possessed larger farms, but questions of land distribution were not pursued. Opinion leaders were analyzed only as communicators, and their possible socio-political roles in the community were not explored. The fact that innovators were consistently characterized as possessing more education, more resources and more land failed to suggest that the innovator group may have been a rural elite attempting to maintain its privileged position by means of various organizational and institutional arrangements. Finally, the diffusionist perspective was based on the assumption that information, and therefore technological change, would trickle down from more innovative to less innovative persons, and that the change agent could stimulate this process through contact with opinion leaders. This permitted the change agent to organize his work with a small group of progressive individuals, relying on an assumed multiplier effect to carry the innovation to poorer and more disadvantaged persons.

The role of the behavioral scientist within the diffusion approach to development has been largely (seen as) that of facilitator responsible for the rapid adoption of innovative practices. He has organized communications networks, trained communicators or change agents, and monitored the innovation to determine its degree of social acceptability. He occasionally has been asked to evaluate the success of the program as determined by its extent of use in the local society, rather than by its distributional impact on social welfare or other success indications. [Saint and Coward, 1977:733–734]

The diffusion studies also suffered from the limitations of focusing on one particular innovation at a time. Data gathering was greatly simplified, but the false assumption was imposed upon the research design that the adoption of any single innovation would yield marked benefits to the adopter. As is now recognized (Mosher, 1966, 1969), it is very rare in agriculture to find that the adoption of any single innovation, leaving everything else constant, will yield marked benefits. Improvement in the economic situation of the small farmer generally depends upon the fit of any innovation with the other elements making up a farming system.

This realization guided the thinking of those who turned to introducing change through a package strategy. That is, the professionals put together a set of inputs and practices that they thought necessary to produce substantial yield improvements and then tried to persuade farmers to accept the package.

Although this approach was more realistic than the single variable emphasis, it, too, had deficiencies. In the early years of agricultural research, extension agents as well as researchers tended to underestimate the variability in climate, soil, and water that conditions the success of any package program. A package might work very well at a particular location and poorly at another not far away. The researcher could not assume that a package would always produce the expected results unless it had been tested and adapted locally to match conditions prevailing at the point of intervention.

Community Development. In the 1940s and 1950s there was another popular line of research and theory, which was all too compatible with the diffusionist research model. This was the era when community development was widely expected to be the answer to problems of rural economic change and progress (Blair, 1982). Practitioners and theorists alike assumed that progress must come through grass-roots participation in the villages and countryside (Cary, 1970). Reacting against the prevailing pattern of bureaucratic organization, community development theorists saw the role of the outside professional or change agent as that of helping villagers to articulate their "felt needs" and then to organize them to work to meet those needs.

Community development was based upon one sound idea: that rural people's active participation in the development process was essential for their economic and social progress. The value of this idea was unfortunately lost because it treated social relations in an economic and technological vacuum. That is, restructuring of social relations was seen as the answer to all problems, regardless of the activities being carried out by the farmers or of the socioeconomic structure and distribution of political power in the community or area.

To the extent that intergroup conflict in a village was based upon poor communication and misunderstandings, a skillful community organizer might help to develop social processes that would advance people toward shared goals. But what of the very common situation in which large and small landholders have opposing interests or there is a conflict based upon access or lack of access to irrigation water?

The socioeconomic structure and distribution of political power in the community may make the difference between whether the change agent can or cannot do effective community organization work. If he enters a village that is highly homogeneous in size of landholdings and in agricultural activity, the change agent with social skills and general background knowledge may be able to help the community members to improve conditions. On the other hand, suppose he approaches a situation at the opposite extreme in the distribution of wealth and power. If power is concentrated in the hands of a small elite that derives important advantages from the poverty and powerlessness of the other citizens, the change agent can hardly expect to help poor farmers through a participation and communication strategy designed to arrive at communitywide consensus.

As Oscar Núñez del Prado (1975) recognized when he intervened in the Indian community of Kuyo Chico in the Peruvian Highlands, progress for the Indians depended upon breaking the domination over them by the mestizo elite in the district capital. Besides helping the Indians to organize, Núñez del Prado provoked a conflict with the mestizo authorities in a situation that allowed him to claim the support of powerful individuals and organizations in the central government.

When conflicts of interest are based upon major cleavages, strategies for development must be adapted to these structural conditions. Whether we are concerned with accelerating the adoption of a particular innovation or, more broadly, with advancing community development, we must devise strategies that fit the social, economic, and political structures of community, region, and nation.

The limitations of the diffusion and the community development approaches became increasingly apparent when technical assistance programs, based explicitly or implicitly on the diffusion model, failed to generate either rapid modernization of the traditional farms or rapid growth in agricultural output (Hayami and Ruttan, 1971:39).

The High-Payoff Input Model. The inadequacy of policies based on the diffusion model led in the 1960s to a reexamination of the assumptions regarding the availability of a body of agricultural technology that could be readily diffused from the developed to the developing countries (Hayami and Ruttan, 1971:39–40). Economists argued on the basis of their research findings that only limited productivity gains could be had by the reallocation of resources in traditional peasant agriculture. Theodore W. Schultz (1964) set forth this position in his book *Transforming Traditional Agriculture,* arguing that peasants were rational but remained poor because they lacked the technical and economic opportunities to which they could respond. Given the appropriate technology and modern inputs, the traditional peasant would respond.

The utility of the high-payoff input model seemed to be borne out by the early success in the development and spread of the new fertilizer-responsive varieties of wheat and rice. Hayami and Ruttan (1971:42), however, argue that this model remains incomplete as a theory of agricultural development because it ignores the mechanisms by which resources are allocated among research and other private and public sector economic activities. Furthermore, it is more a model of growth than development in that it fails to address a major theme of this book, the issue of equity in agricultural development.

Agricultural Sector Studies. An important segment of the agricultural economics profession in the 1960s was concerned not with farm-level research but with macro agricultural sector studies, which were a major priority of USAID-funded social science research. Computerized agricultural sector models were developed on the argument that the complexity of the agricultural sector made mental arithmetic and intuition inadequate bases for the identification of bottlenecks and comparison of alternatives (Rice and Glaeser, 1972:25). The Michigan State University simulation models of Nigeria and Korea, the Iowa State University programming model in Thailand, and the World Bank programming model in Mexico are among the better-known examples.

In retrospect these models were very costly to build and maintain both in money and, even more important, trained manpower. Furthermore, most of the analysis proved to be of minimum utility to planners and policy makers. John L. Dillon (1971:85) has promulgated three laws of simulation which help us to comprehend the reasons for the failure of this approach:

(1) Simulation like statistics cannot prove anything.

(2) Simulation like statistics can nearly prove anything.

(3) Once started simulation will continue until available funds are exhausted.

The funds for sector analysis studies were exhausted in the 1970s when funding for research was sharply cut and new priorities were established.

Controversy over the Benefits of the Green Revolution

The controversy surrounding the interpretation of the consequences of the Green Revolution has helped to redefine the role of social science research in a more positive light. Those who developed the initial technology of the Green Revolution, being principally biological scientists, gave little consideration to the socioeconomic implications of their work. New technology was considered essential to raising production because the land frontier was closing and the man-land ratio was increasing rapidly in most of the developing countries of Asia. As suggested by the American and Japanese experience, continuous technological innovation in seeds, inputs, and implements was seen as the cutting edge of the agricultural transformation (Hayami and Ruttan, 1971). The distinct difference in the historical patterns of development in countries with resource endowments as different as Japan and the United States were not well understood.

The new seed-fertilizer technology was developed in the experiment stations favored by fertile soils, well-controlled water, and other factors suitable for high production. There was little perception of the complexity and diversity of the physical, let alone the social, environment in the developing world. In retrospect this proved to be a mixed blessing. If the modest research resources for the international agricultural research systems in the 1960s had been concentrated on the less favorable environments, no major breakthrough would have been made. The scientist, however, frequently saw his job and responsibility as ending at the experiment station gate. The failure of farmers in many areas to adopt the modern technology was attributed initially to peasant conservatism and backwardness and to the failure of extension to do its job in disseminating the technology.

Concern over the failure of farmers to adapt to the new technology is reflected in the statement of the former director of the International Rice Research Institute, Robert F. Chandler, Jr., who wrote in 1978:

On retiring from IRRI in 1972, the only real disappointment I felt was that somehow we did not understand sufficiently why the Asian farmer who had adopted the new variety was not doing better. Somehow I felt that rice scientists who had obtained yields of 5 to 10 metric tons per hectare on the IRRI farm still could not explain why so many Filipino farmers (for example) obtained on the average less than one metric ton increase in yield after shifting from traditional to high-yielding varieties. All of us were a bit mystified as to why not more than 25 percent of all rice land in the less developed countries was planted to the new varieties. [Chandler, 1975:15]

As the cereal grain technology spread, social scientists developed their own interpretation of events. Two very contrasting points of view arose, which we can broadly classify as induced innovation and the dependency theory. These contrasting positions are reflected in many of the articles contained in *Science Politics and the Agricultural Revolution in Asia,* the proceedings of a conference of social scientists sponsored by the American Association for Advancement of Science (Anderson, Brass, Levy, and Morrison, 1982).

Advocates of the induced innovation model saw the course of events, as Hayami (1982) has argued, in this way. The situation in developing countries resembles the world of such classical economists as David Ricardo. As the growth of population presses hard on limited land resources under constant technology, cultivation frontiers are expanded to more marginal areas and greater amounts of labor are applied per unit of cultivated land; the cost of food production increases, and food prices rise. In the long run, laborers' income will be lowered to a subsistence minimum hardly sufficient to maintain stationary population, and all the surplus will be captured by the landlord in the form of increased land rent.

The Green Revolution was seen as a response to these circumstances. Rising food prices and declining fertilizer prices made it increasingly profitable to develop new fertilizer-responsive varieties of cereal grains (induced technological change). New technology was seen as the mechanism to escape the Ricardian trap and the inequities that would result from a lack of growth.

Whether the technology-induced growth would lead to greater equity was a moot point. At one extreme was the argument that if free market prices were allowed to prevail and reflect the true social value of resources, development problems would take care of themselves (price purists) or, alternatively, the increased wealth could be redistributed through a welfare program (see, for example, Schultz, 1978, and Owen, 1971).

A somewhat different view, reflecting its own variety of technological determinism, held that whether the new technology promoted equity or reinforced inequity was determined by the nature of the technology itself and by the institutional setting. In attempting to develop new technology that promotes greater equity, three issues are important: the choice of crop, the choice of environment, and factor bias (Dillon, 1979). Work in staple foods that benefit the poor in disadvantaged environments and on labor-using and land-saving technologies is seen as promoting equity. The overemphasis on irrigated areas is seen as having created a disequilibrium. But whether research should be concentrated on developing technology for the disadvantaged environments, or, alternatively, whether priority should be given to research for those areas with greater potential for technological change has been a matter of debate in developed as well as developing countries. The appropriate answer will depend, of course, upon analysis of the social benefits and costs for a specific situation.

The early literature of the Green Revolution reflected a polarization of views between the supporters and detractors. Scholars studying the same events, and in some cases the same data sets, drew opposite conclusions about the impact on equity. This dialectical debate has led to a greater understanding of the complexity of the issues for social and biological scientists alike and has encouraged a reevaluation and redirection of research emphasis.

One consequence has been the greater effort to develop technology suited for the less favorable agroclimatic areas. The need for a more holistic look at the farm family as a decision-making unit and for a greater interaction between the farmer-client and the researcher is reflected in the development of cropping systems and farming systems analysis. This interactive approach to research, discussed in Chapter 12, is still in the formative stage. The effort is focused largely on unirrigated environments, where the problems are more complex and the impact of research on productivity and equity is as yet uncertain.

Dependency theory, first popularized in Latin America and now widely recognized around the world, particularly by those with a Marxist orientation, accepts the primacy of structural conditions and focuses upon the distribution of power (Kahl, 1976). Furthermore, the theorists have emphasized the importance of the distribution of power at the national and international levels, paying less attention to power questions in the village.

The dependency theorists argue that the new technology has widened the gap between the rich and the poor and that this result was no accident. The center has gained at the expense of the periphery; that is, the developed countries gained at the expense of the developing, those in power in the LDCs gained at the expense of the peasants, the large farmers at the expense of the small, the landlords at the expense of the tenants. The theory that benefits would "trickle down" to the poor had not worked and perhaps was never intended to work. The Green Revolution had strengthened the hand of those in power. Without major institutional reforms, efforts to introduce new technology were misspent. Furthermore, there was an inherent bias in the exotic technology of the international centers in favor of the rich. Technologies that will help resource-poor peasants must be developed by the indigenous scientific community working hand in hand with local farmers.

Structural Change Theorists. Without passing judgment on the Marxist thesis of the necessity of a national revolution, many social scientists have become convinced of the importance of structural conditions (the socioeconomic structure of the community, the distribution of political power, and the economic activities carried on by the inhabitants) as the primary factors in promoting or blocking economic and social progress for the poorer segments of the community. Those whom we may call structural change theorists consider economic and technological factors to be necessary but not sufficient for development. Though not uninterested in psychological orientations, they assume that these orientations will change as objective conditions allow poor farmers to find new opportunities for improving their condition. In other words, they assume that the poverty of the rural community is based on its disadvantageous position in relation to its physical, economic, and political environment. This assumption suggests a need to concentrate on the relations of the small farmers to their environment and, in practical terms, to devise ways in which the environment could change so as to provide the farmers with new options likely to improve their welfare.

There is also an important contrast in implicit assumptions about the role of change agents as we move from diffusion to structural change approaches. Diffusion theorists had been inclined to see progress as being caused or at least accelerated by the intervention in traditional communities of change agents. Ideally, such individuals would have some knowledge of, and respect for, the local culture. But their essential role was to determine what small farmers should do to better their lot and then to devise strategies to overcome resistance to change. Structural change theorists have tended to minimize the importance of individual change agents and have instead concentrated upon carrying out structural changes that will open up new possibilities for small farmers to seize new options and thereby make progress.

Although diametrically opposed in their diagnosis of the problems of the poor, diffusion and dependency (Marxist) theorists share the assumption that change must be brought about by change agents from outside the community. The difference is that diffusion theorists see the change agent as functioning to alter psychological orientations of peasants so that they will progress more rapidly within the existing structure, whereas dependency theorists look on change agents as activists who will mobilize poor people and indoctrinate them on the necessity to act collectively toward making revolution. In general, dependency theorists have shown little interest in studying indigenously created changes and have seen little value in local area studies.

Those pursuing a structural approach, whether or not they embrace Marxism, have provided useful stimuli, yet their formulations are of limited value at the grass-roots level. On the constructive side, they have provided a useful corrective against treating development problems as essentially psychological, warning us that if we treat human relations and communications in a structural, economic, and technological vacuum, we can make little progress with such limited comprehension of impediments to change. On the other hand, the recognition that structural elements are of great importance simply points to a problem area and fails to provide guidance on what government officials and independent professionals can do directly to improve the lot of small farmers in poor communities. In fact, most Marxist theorists simply decline to address this problem, arguing that nothing very useful can be done at the local level until major structural changes are carried out regionally and nationally.

Marxists also seldom address problems of changing bureaucratic organizations. In the past, they have tended to look upon government bureaucracies as evil and standing in the way of progress. Thus they have not considered how a better understanding of those organizations might provide ways to make them more responsive to the needs of the poor rural majority. It is unrealistic to assume that there is no need to understand current bureaucracies because even after a revolution, government structures and processes would not be completely changed. Large-scale bureaucracies would continue to exist and to shape much of what happens in society. There may be differences in the way a government bureaucracy functions before and after a revolution, but research so far suggests that bureaucratic organizations have many features in common, whether the economic systems they represent are labeled free enterprise, communist, or socialist.

A Social Science Framework for Agricultural Research and Development

It is always easier to point out deficiencies than to present a framework to provide a sounder basis for research and action, but let us try to begin this more difficult task. The framework we see emerging focuses on the social organization of agriculture. Here we interpret "social" broadly to encompass variables traditionally dealt with in sociology, anthropology, psychology, economics, and political science. In other words, we need an interdisciplinary framework so that, as specialists in different disciplines work together, they understand how their particular data and interpretations relate to the data and analyses of specialists in other disciplines. This social science framework must be linked with the technology of agriculture and with the conditions of the natural environment within which farmers live and work.

This framework would cover at least the following three elements: factors influencing the motivation of small farmers, particularly as they make decisions in their agricultural activities; the nature of development projects and change processes; and community and bureaucratic organizational structures and social processes.Elsewhere in this book we have concentrated upon this last category. Without claiming that we have accomplished more than a preliminary treatment of this important area, let us concentrate on the other two main elements which we have not explicitly treated before.

The Motivation of Small Farmers in Decision Making. We assume that the desire to improve one's economic condition is almost universal, but at the same time we must recognize that this motivation operates within a social framework. Here the approaches of economists and other social scientists should complement each other. In assuming that human beings guide their behavior to a large extent on the basis of past consequences and future anticipations of the material costs and benefits of particular actions, economists do not claim to be able to predict the behavior of particular individuals or groups of people. They simply assume that aggregates of humans will act in such ways that one's understanding of behavior will be enhanced by assessing the potential costs and benefits associated with that behavior. Economists, of course, recognize that noneconomic factors influence behavior, but their discipline does not provide theoretical tools for dealing with these other factors.

In emphasizing the importance of noneconomic factors, other social scientists have at times underrated the importance of economic factors. Of course, it is fruitless to argue whether economic or noneconomic factors are more important. For our purposes, we cannot use a theory of economic man or a theory of noneconomic man; we need to have a theory of socioeconomic man, which requires placing the individual in the context of family, community, and organizational memberships and relationships.

Under some conditions, it is useful to assume that particular individuals respond to their assessments of the anticipated costs and benefits of contemplated actions. The following seem to us the limiting conditions.

(1) It is an individual alone who decides to carry out a certain line of action. The decision is not made for him or her by others.

(2) The individual is free to choose among several options.

(3) Implementing the decision does not depend upon the conscious collaboration of others.

(4) The individual himself will pay the costs incurred and will receive the benefits. Under these conditions, and if reasonable estimates can be made of the costs and benefits expected from a given decision, the researcher is in a good position to explain or predict the behavior of the decision maker.

A fifth condition might be added: the individual has full knowledge of the costs and benefits. This condition seems unrealistic and unnecessary. Human beings rarely have full knowledge in advance of the costs and benefits of a particular action. As a rule, we are not seeking to explain a once-in-a-lifetime decision, for which past experience provides no guidance. We assume that the individual makes judgments on the basis of the outcomes of decisions made in similar situations in the past. He makes mistakes because his judgment is imperfect or because the future is different from the past. When the individual finds he has made a mistake—that is, when the costs prove to be much greater and the benefits much less than expected—he takes the information derived from this new experience into account when making the next decision. The individual never has complete knowledge of costs and benefits before deciding but is constantly reassessing expectations regarding outcomes on the basis of experience.

As other social scientists are quick to point out (and as economists recognize), human beings also seek nonmaterial benefits. This customary objection is not as serious as sometimes thought, however, because humans, whether peasants or professors, are almost universally concerned with material costs and benefits.

The more serious problem involves the individualistic nature of the formulation, completely disregarding social and organizational influences on decision making. As previous chapters have argued, a realistic model of small farmer decision making does not start with the individual farmer but rather with the farm family. The farmer thinks and acts in the context of the household economy. He (or she) must consider any possible innovation in terms of the costs and benefits to the family, balancing increased costs in money and family labor (or hired labor) against anticipated increased income—and always keeping in mind opportunity costs: what family members might be able to earn off the farm if they did not contribute the extra labor that the improvement requires.

The family is also embedded in a community social system involving the farmer in a reciprocity network. Let us say, for example, that it is now the turn of Farmer X to serve as *mayordomo* for the annual ceremony in honor of the patron saint of his village. In this village (as in many others) the expenses incurred by the *mayordomo* are so heavy as to constitute a serious drain on the family resources, force him to sell some animals, and go into debt. If the farmer were free to choose, he might well prefer to put his resources into an innovation to improve his farm. In fact, he may confess to the visiting professional that he wishes he did not have to accept the ceremonial obligations. But, having depended on others in the past to shoulder the financial burdens of the saint's day, Farmer X cannot escape his turn without antagonizing relatives, friends, and fellow villagers, with whom he must continue to live and from whom he may need to get labor or financial resources for his agricultural production operations from time to time.

In such a situation, it will be futile for the outsider to try to persuade the individual to put money into the farm instead of into the ceremony. Such a shift in resources must be viewed as a collective rather than an individual decision-making problem. As the burden of ceremonial expenses becomes generally recognized throughout the community, an outsider may help the villagers to make that shift through agreeing to reduce the costs of the ceremony and/or meeting expenses collectively, rather than placing them primarily upon one individual at a time. In other words, the farmer's freedom to allocate more of the family resources to improving the farm may depend upon a prior collective decision to ease the financial burden on the *mayordomo*.

We must also go beyond the diffuse influences of social and cultural patterns to consider the relationship of the individual to organizations established to seek certain objectives. When the individual cannot pursue a line of action unless others act with him and his family, analysis of behavior must go beyond the simple estimating of potential costs and benefits in order to assess the organizational requirements for success in this task and the perceived probabilities that others will not only agree to a given decision but also fulfill their commitments.

To provide a firmer basis for analysis of the farmer's motives, we must consider both priorities in the benefits sought and organizational requirements for efficient action. We must recognize that improving farm yields and farm income may not always be the number one priority of the small farmer—as agricultural development professionals are likely to assume. Only after Caqueza professionals had helped the farmers to make contact with the government electric company was it possible to capture villager interest in improving agricultural technology.

If the farmer's success depends upon actions of other people, we must consider the organizational requirements for efficient action: what people are to be involved and how their collaboration is to be secured.

Weighing Risks

No discussion of motivation in decision making would be complete without considering risks. Like other people, in deciding upon a future action, small farmers must consider the costs incurred and the benefits secured from similar actions in the past. Even for familiar actions, farmers recognize risks: the costs may be higher and the benefits lower than they have anticipated. When farmers contemplate a line of action without parallels in their experience, their weighing of risks tends to assume overriding importance.

It is important to recognize that risk comes in various forms, at least three of which need to be distinguished for our purposes: environmental, local social, and institutional.

Environmental risks involve sudden and drastic adverse changes in climate (droughts or floods, for example) or in prices for crops in the farmers' market. Farmers cannot affect the climate, and small farmers generally have little power to affect market prices. Except for long-range projects (a massive reforestation project, for instance), governments are powerless to affect the climate. Governments can and do affect prices either by seeking to hold down prices to benefit urban consumers or by making government purchases at support prices when market prices fall below a certain level.

Without being able to control environmental risks, small farmers necessarily take them into account in accepting or rejecting proposed improvements. For example, farmers may be persuaded that a new plant variety will outyield the native varieties under average conditions, but they also want to know how the high-yielding variety will fare under drought conditions. If their area is subject to droughts and if they believe (rightly or wrongly) that the high-yielding variety will not withstand the rigors of a drought as well as their native varieties, they are unlikely to make the change.

Collective projects involve an element of local social risk. When the project depends upon the cooperation of a group of people, the individual may put in the labor and capital expected and yet not gain anticipated benefits because others do not meet their obligations. In such a case, the decision maker will not only seek to judge whether, if the plan is carried out, the objective will be achieved. He also has to estimate the probability that his fellows will meet their commitments. Failure of cooperation is an important risk that must be, and, in fact is, commonly taken into account by small farmers.

In a series of studies carried out in Peru, the United States, and the Basque country of Spain, we have found a wide range of differences from community to community and culture to culture in the tendency of people to trust each other and have confidence in the behavior they can expect from their fellows (Whyte, 1964; Williams, Whyte, and Green, 1966; Johnson, 1976). But, we should not assume that these cultural characteristics are immutable. After a series of unsuccessful experiences, individuals will see great risks in attempting any new collective effort. Conversely, a series of successes in cooperative projects builds social capacities, which have indirect but very important economic value. By raising the perceived probability of success through collective efforts, these past successes encourage people to make further investments of their time, energy, and resources.

The authors of the Caqueza study suggest an additional category which they term institutional risk (Zandstra, Swanberg, Zulberti, and Nestel, 1979). Success of a project may depend upon actions of people in organizations outside of the community. For example, the success of a given innovation may depend upon the availability of the required inputs in a market accessible to the farmers, the delivery of credit at the time needed for best results, and the presence of an agricultural professional in the village to give technical assistance when the innovation is first tried. It is possible to devise crop insurance schemes based upon studies of the frequency and severity of droughts, floods, and other natural disasters, but no government has been able to provide insurance against institutional risks. Lacking such protection, the farmer may well decide that the probability of joint occurrence of the three conditions for success is extremely low. In such a case, the rational decision is to reject the innovation. It is then useless for change agents to concentrate efforts on persuading the farmers to adopt the innovation. Success may depend upon devising ways of reducing institutional risks.

There is another institutional risk generally overlooked by students of agricultural development. Villagers may expend much time and money in trying (often unsuccessfully) to get the government to deliver what are supposedly free services. Consider the following case.

Instead of waiting for the national government to provide this service, the village of Huayopampa (population 471) in the Peruvian highlands hired an engineer to draw up plans for a local public works project. There were two reasons for this unusual act of self-financing. Having shifted from traditional crops to commercial fruit growing, Huayopampa had become more affluent than the average village and, as a village leader explained, the people had learned from years of experience in dealing with the government.

> Over a period of fifteen years we sent delegation after delegation to Lima to get the government to provide us with the material and the technical knowledge we needed to carry out the project. This was a very frustrating experience because we were always getting promises and no action. Besides, when we came back and nothing happened, there were always people in the community saying that we had just spent the money to have a good time in Lima. Finally, after fifteen years the government did provide the material and the engineering help we needed so we could complete the project. But then we sat down and figured out how much it cost us in expenses for those trips to Lima over the years, and we realized that we would have got the job done much faster and cheaper if we just had bought the materials and hired the engineers ourselves.   [Whyte and Alberti, 1976:177]

In Peru, as in many developing countries, local communities have very limited abilities to tax and therefore tend to be heavily dependent on the national government to carry out major improvement projects. Poor as most of them were, we found villages customarily including in their annual budgets an item called "costs of representation." This money financed villagers on their trips to Lima to seek government help and covered expenses for entertaining national officials when they visited the village.

In a poor country, there is always much more demand for services and financing than the government can possibly meet. Securing government services is therefore a competitive process, with each village trying to outdo the others. Our informant correctly observed that securing these "free services" had cost Huayopampa a great deal of money.

There may be other advantages for a village in avoiding dependence upon some government services. In a government project, the national officials are inclined to view their tasks as favors to the local people, who therefore have no right to insist on who should do the work or when and how it should be done. When the village government or cooperative organization finances the project, local officials have more power to choose who shall do the work and how and when. Those hired are accountable locally, which can make a great deal of difference in the efficiency and timeliness of the project.

The importance in development planning of devising means whereby villagers gain some continuing control over the flow of money is clear. Furthermore, this control must be collective for, in a community of small farmers, few families have extra money to spend. If they are to pool their resources to meet the costs of community projects, they must build a local organization capable of raising and disbursing money. To the extent that they can minimize their dependence on the national government, the villagers can minimize institutional risk.

Organizational Requirements for Improvement Projects

In understanding farmer behavior, we need to go beyond assessments of costs and benefits and of risks, nor can we be content when we have simply placed the decision-making process in the social context of family and community. We must also examine the organizational requirements for the successful implementation of the project being considered. If we do not take this step, we may fall into the trap of assuming that a project idea is rejected because of resistance to change rather than recognizing that the organizational requirements for successful implementation were beyond the capacities of the farmers considering the idea. In weighing any new project, therefore, we need to consider not only the farmers' decision-making process but also the organizational requirements for successful implementation.

We need to get down to cases and ask ourselves for each case, first, what are the potential costs and benefits influencing the motivation of the parties concerned and, second, how does the nature of the project tend to shape the socioeconomic requirements for its implementation. In the previous sections, we presented a framework for analyzing farmer motivation. It has been impossible to deal with the motivation question without some considerations of the nature of the project. We now present a more systematic framework for analyzing the nature of development efforts.

Change projects open to small farmers are not uniform in character but are varied, requiring different structural conditions in the organization of activities and in the distribution of costs and benefits. We have identified four types, each of which has different organizational requirements and a different pattern in the distribution of costs and benefits (Whyte and Williams, 1968). These are as follows:

Individual-Direct. The individual makes the decision and he or his family pays all of the costs and receives all of the benefits. Examples are the introduction of a new high-yielding variety, a new method of cultivation, a new chemical fertilizer or pesticide—providing the farmer can make the change without securing the approval or collaboration of fellow villagers. Here the organizational requirements involve only the coordination of the farmer's actions with those of other family members.

Individual through Group, with Equitable Sharing of Costs and Benefits. The individual decides whether to participate in the action, but the project cannot be carried out unless other members of the group also participate. Examples are building or improving an irrigation system or building or improving a road to give the village better access to the market. In considering such a project, villagers will seek some assurance from each other in advance that the sharing will be equitable—that if some incur greater costs or gain greater benefits than others, they do so as a result of circumstances beyond the control of project leaders and not of some individuals taking advantage of others.

Unequal Sharing of Costs and Benefits. Here, the nature of the project is such that its costs will fall more heavily on some people than others or the benefits will be shared unequally throughout the community. For example, a reforestation project may provide equal benefits to those who are paid for planting the trees but reduce the land available for cattle raising. This reduction may be of no consequence to families with few animals but a major loss to those with large herds.

Control of Individual Interests in Order to Maintain or Increase Group Benefits. This type of project is frequently found in cattle-raising communities where it is to the advantage of each family to have a maximum number of animals but where such unrestrained individualism would result in overgrazing, turning the range into a desert.

The outsider who wishes to facilitate the development process must help villagers take into account the possible fit between the current social system of the community and the organizational requirements of the project they are considering. Let us illustrate this point by relating the organizational requirements of each type to the community social system.

For type 1 (individual-direct), the organizational requirements are the least difficult. The individual family can adopt the innovation without cooperation from any others. The family incurs the costs and also receives the benefits. The outsider can easily be misled by not recognizing possible social system or technical system constraints. As an example of social system constraints, we have already considered the ceremonial obligations of the *mayordomo* which may make it impossible to devote the resources needed. An example of a technical constraint would be the case of introduction of a new high-yielding plant variety, whose success requires a change in the scheduling of water distribution through the community irrigation system. It is futile to try to persuade the farmer to change when success depends upon a community decision to change the water distribution pattern.

Type 2 (individual through group, with equitable sharing of costs and benefits) offers attractive possibilities of building community solidarity. If few such projects have previously been carried out and villagers have little faith in their ability to work together, the outside facilitator should encourage villagers to begin with a project having the following characteristics: planning allocation of costs and benefits is relatively simple; the benefits are forthcoming in a short time after the costs have been incurred; the benefits promise to outweigh the costs substantially; and it is not difficult to withhold benefits from those who fail to cooperate.

The last point is less familiar than the others and so is likely to be overlooked. The enforceability of the agreed-upon allocation of costs and benefits depends upon the strength of the local government or other association and the nature of the project. As an example of the first point, consider the cases of two Peruvian highland villages, both of which had norms requiring the levying of fines upon farmers who did not meet their communal obligations. We found that in Huayopampa, 100 percent of the fines were collected, whereas Pacaraos was able to collect less than one-third of the fines (Whyte and Alberti, 1976). Clearly, planners for Pacaraos need to give special attention to devising projects whose nature facilitates enforcement. For example, in the building or improvement of an irrigation system, it may be possible to cut off the water from families who fail to contribute until they pay a fine. Similarly, when an electric lighting system is constructed, a noncontributing family can be denied a hookup. On the other hand, there is no practical way of denying noncontributing families use of a communally built road.

For type 3 (unequal sharing of costs and benefits), the outside facilitator needs to know not only which families stand to gain and which to lose but also the relative positions in the social structure of the gainers and losers. Inevitably, such a project involves a conflict of interests, which cannot be resolved simply by trying to stimulate good communication. The ultimate outcome of the project will not necessarily leave some people with net gains and others with net losses. Those planning the project can stimulate a discussion to make explicit the potential gains and losses and to devise ways in which the potential losers can be compensated. The process may involve bargaining rather than discussion leading to a community wide consensus. A type 3 project is likely to be a good deal more difficult to implement than types 1 or 2. It is especially likely to require strong local government or community leadership, perhaps reinforced by outside technical assistance.

Type 4 (control of individual interests in order to maintain or increase group benefits) is likely to be the most problematic since it involves restraining individual families from doing what is in their immediate interest. Probably there are few villages with local governments or cooperative organizations strong enough to implement such projects without outside help. The community may need not only technical assistance but also help in enforcement from provincial or national government. A villager assessment of institutional risk—an estimate of the probability that the government will deliver the enforcement measures promised—will be needed.

An Agenda for Applied Social Science

Social scientists interested in action as well as knowledge should develop a participatory style of community involvement. If the research is done in the traditional style of experts, villagers are unlikely to understand the findings or to accept their implications. And the social scientist can hardly be effective in facilitating change if, as the project moves from diagnosis to action, he abandons the participatory approach and reverts to the traditional role of expert, telling people what to do.

Progress in both research and action depends in part upon changing the role of the outsider and even the title we use for him. The titles currently in common use—expert, consultant, and change agent—all tend to suggest a person of superior wisdom who knows (or can readily determine) what the villagers should do and who has the responsibility of getting them to do it. Although such terms can be reinterpreted so as to downplay their authoritarian implications, we prefer to use the term of *facilitator*. The facilitator is an outsider who comes into the community or organization in order to help the members (and himself) understand their conditions, diagnose their problems, devise new options for action, and decide what they themselves want to do.

Social scientists should participate in field studies of farming systems, but their distinctive contributions would continue to be studies of the socioeconomic structure. Attention should focus on the nature of the household economy, on the supply and demand for labor, and on the customary division of labor by sex and age. Social scientists should be particularly concerned with the land tenure system: the distribution of landholdings by size, access or nonaccess to irrigation, and patterns of owner-tenant or owner-sharecropper relationships. In some cases, there may be additional relations between owners or tenants and landless people involving labor requirements at peak times (planting and harvesting) to get a share of the crop (Kikuchi and Hayami, 1980). While giving special attention to family income and expenditures, social scientists should recognize that these hard data need to be interpreted within a context of the local culture. A farmer's ceremonial obligations may make it impossible for him to expend the money needed to carry out a farm improvement project.

Project planners should be particularly interested in farmers, but they must also look at the total occupational structure of the community, since in some cases schoolteachers or other professionals, merchants, or government employees may play important roles in facilitating or impeding the progress of an agricultural project. The social scientist should be concerned with the presence or absence of organizations (cooperatives, communal work groups, and the like) which may support or oppose a change project (see particularly Esman and Uphoff, 1982). Finally, such socioeconomic studies should dwell upon the structure and functioning of local government and upon the distribution of power. Here researchers should be concerned not only with the formal organization of government but also with its capacity to facilitate or impede change projects, as reflected in past experience.

An intensive scientific study of all of these aspects of the community could require a year of the time of one or more researchers. It is important to distinguish between such a project and what we call a diagnostic study (Bradfield, 1980). If the purpose of research is to produce a community study that will be a contribution to the scientific literature, then a major investment of time and resources will be required. If the purpose is to guide an agricultural project, it is possible to develop shortcut methods that enable researchers to carry out a study in much less time.

In the first place, researchers rarely need to start from a zero base of knowledge. From previous written reports or from people with local experience, researchers can gain a general picture of the area, which will help to guide them from the known to the unknown. Researchers will know in advance whether the community is characterized by a few large landholdings and many very small holdings or whether it is a relatively homogeneous community of small owners. More precise estimates of the size and distribution of holdings will require field work.

The choice is not limited to either a scientific study of a year or more or a two-week diagnostic study. As we have noted in Chapter 17, how much should be invested in agrosocioeconomic research should depend upon the problems and progress of agronomic and animal husbandry experimentation. In an area that appears to be similar to sites of previous research and experimentation, a two-week agrosocioeconomic diagnostic study may be adequate for a beginning. Later, if projects are not proceeding satisfactorily, the situation may call for more intensive research in the plant and animal sciences and in the social sciences as well.

Villagers themselves will be the primary sources of information in such field studies, but that is only a necessary and not a sufficient condition for establishing a participatory research process. Playing the role of expert, the researcher can treat villagers as passive subjects that he pumps for information on which to base analysis and conclusions. In the facilitator role, the researcher establishes a collaborative relationship with the villagers, consults with key people in planning the study, asks not only for information but also for interpretations of its meaning, keeps people informed on the progress of the study, and asks their advice on next steps. The researcher may even find it advantageous to recruit one or more villagers to play the roles of paraprofessionals, collaborating in the research process. From time to time, the facilitator feeds back to the villagers, in a forum they help him to arrange, progress reports of the study. Such feedback serves not only to build rapport between the researcher and the people he studies. Feedback sessions can also improve the scientific quality of the research as villagers point out errors of fact and alternative explanations for findings (Whyte, 1979; Fortmann, 1982).

Such a collaborative research process should flow naturally into the action phases of the project. In the action phase, the role of the facilitator is sometimes interpreted as that of a resource person, but that term leaves it unclear when and how the resources are to be brought into play. Generally, facilitators enter the field armed with a body of specialized knowledge and skills derived from an academic discipline. If facilitators do not impose this specialized knowledge, how can they contribute the resources they bring to the village?

The rhetoric of community development stresses the importance of responding to the "felt needs" of the community. But suppose the villagers do not articulate any felt needs requiring the specialized knowledge of the outsider? In this situation, specialists are likely to try to create the felt needs that will call for their specialties. When specialists have had some exposure to community development doctrines but have not advanced beyond the initial cliches, we may find a public health professional trying to create felt needs for health improvement, an engineer trying to create felt needs for a public works project, and an agricultural specialist trying to create felt needs for better plant varieties or better methods of cultivation.

Disciplinary competition is wasteful of the scarce resources of specialized knowledge potentially available to the community. In trying to create a felt need, the change agent is at the same time selling his or her brand of specialized knowledge and, in determining the direction of change, perpetuating the dependence of villagers upon outsiders, which is a major obstacle to self-sustaining local development.

What are facilitators to do if the needs that their discipline equips them to deal with prove not to be high on the list of village priorities? In the first place, the answer can be anticipated in part by the selection of the communities to which facilitators make their approach. One whose knowledge, skills, and interests are in the field of agricultural research and development does not voluntarily choose to work in a community where agricultural activities are unimportant. On the other hand, even where agriculture provides the main economic and social base for the community, the facilitator cannot be certain that any given project in agricultural improvement will be the number one priority for the villagers.

In the rural electrification case, the Caqueza staff members responded skillfully, avoiding two potential errors. On the one hand, they did not try to persuade the villagers that they really needed agricultural improvements more than electric light and power, and on the other hand, they did not assume responsibility for bringing electricity into the village—a project for which they had neither the training nor the organizational position to carry out. Instead, they sought out officials in the rural electrification agency and served as intermediaries between village leaders and these government officials, providing the communication links and the encouragement to make the project possible.

The case illustrates one important resource that the outsider can bring to the village. Villagers often suffer from a lack of political as well as economic resources. They may have had contacts with politicians and government officials at regional levels, but often these will have been sporadic and unproductive. In general, promises of assistance from political or administrative officials have far outrun performance, leaving the villagers pessimistic regarding possibilities of any outside help. The villagers also generally lack information on how particular government agencies function, how responsibilities are distributed among them, who has to be approached for a decision on a given project, and so on.

Through previous education and professional or social experience, the facilitator may have made some government contacts. In any case, he is likely to know a good deal more than the villagers regarding the internal workings of government bureaucracies and so is more likely than they to know where to go and whom to approach on a given project. Finally, superior status and educational background are likely to give the facilitator more confidence in his ability to speak the language of the government official and thus increase the chances of getting a positive response.

The outsider also generally has an advantage over the villagers in geographic mobility. In numerous cases, villagers have made substantial financial sacrifices to get to a state or national capital to seek help from a government official, but doing so can be a heavy burden upon a poor village. In contrast, the outsider will generally be a member of a private or government organization whose main office is located in a regional center or in the national capital. The facilitator will need to make periodic trips there to maintain communication with organizational associates and superiors. On such trips, with little additional cost in time and energy, the facilitator can make the government agency contacts required to link the villagers with an organization appropriate for the project they have in mind.

In playing this role, the outsider may find himself slipping into a position of local leadership because the villagers have depended upon him to establish the initial linkage. To avoid this dependence, it is important for the facilitator to withdraw from active participation in a project not in his own field after the initial linkage. Furthermore, it will be advantageous for the development of local leadership if the facilitator does not make the first contact with a government official alone but is accompanied by local leaders, who thereby gain a more advantageous position and more self-confidence to follow up on the first initiative.

Especially in the early stages of the relationship between outsider and insiders, the facilitator needs to take care to avoid slipping into the role of an authority figure. Through past experience with professionals, villagers may have become accustomed to hearing authoritative statements from outsiders. They may not be persuaded by what they hear, but the difference in social class between villagers and professionals is generally such as to lead the villagers to feign acquiescence rather than to express disagreement. This is their way of being polite to strangers of superior status and also a way of avoiding the risk of antagonizing someone who might possibly do them some harm or good.

To avoid this trap, outsiders should take care in the early stages to make it clear that they did not enter the village with any prepared sets of answers to village problems but are seeking to help villagers to arrive at their own answers. Later, as the parties move toward clarifying options and making decisions, facilitators can express opinions more freely and also bring in information and ideas gained outside the community without fear that the villagers will be overawed by their expertise. One way of testing the strength of the relationship is for the facilitator to ask himself whether and how often villagers disagree with him. Since it is impossible for outsiders and insiders with vastly different backgrounds and experiences to be perfectly in tune with each other, the facilitator who never or hardly ever encounters an openly expressed disagreement should interpret this as a sign of a serious weakness in the relationship.

The facilitator should not try to get villagers simply to make the optimal decision on a particular problem. Villagers in a poor community suffer not only from lack of material resources but also from a lack of power. Poor farmers have grown accustomed to having people of superior social class position come into their community and tell them what they ought to do. One of the major aspects of powerlessness is a lack of choice. In helping villagers to discover new options, the facilitator increases their power over the natural environment and also their power in the social systems of family, community, and nation.

As we have seen in examining the participatory agricultural research programs developing in Guatemala and Honduras, an important aspect of this emerging approach involves the demystification of agricultural research. Problems that are on the frontier of knowledge in the plant, animal, and soil sciences will continue to need the resources of highly trained specialists, but as the agricultural R&D process moves on to adaptive on-farm research involving active participation of the small farmers themselves, social scientists can gain enough knowledge of the biological aspects of agriculture to be able to participate usefully in the on-farm experimental process.

By the same token, more complex socioeconomic problems will continue to demand the best efforts of social research professionals. Plant, animal, and soil scientists can learn enough about culture, social structure, organizational behavior, and the social and economic requirements of particular change projects to work effectively in professional interdisciplinary teams and in projects involving active participation by small farmers.

Behavioral scientists are gaining increasing acceptance in national and international research-extension organizations, but policy makers have assumed that their usefulness is limited to micro-level research or action projects. The common tendency is to see their contribution simply as being sensitive to the needs and interests of small farmers, coupled with an ability to relate well to those farmers in the field.

There are two drawbacks to this conception of behavioral science roles. Undoubtedly there are many traditionally oriented plant scientists who study a farmer's field while minimizing personal contact with the farmer, yet we have encountered many plant scientists who relate well to small farmers and are sensitive to their needs and interests. Thus the assumption of superior social sensitivity is hardly enough to support the broad redefinition of behavioral science roles we now see emerging. We suggest that the bureaucratic reorientation we have been advocating urgently needs social scientists with interests and skills in the analysis of organizational behavior.

A major obstacle to change is a general lack of understanding of the dynamics of government agriculture-related bureaucracies. Since they have had no formal education that might clarify the workings of a government bureaucracy, plant, animal, and soil scientists tend to view the bureaucracy as a black box. You put something into the black box, and much later, something may—or may not—emerge. If something emerges, it may be hard to understand how it is related to your input. What goes on inside the black box remains a mystery.

This black box orientation leads agricultural professionals to view a bureaucracy mainly as an obstacle blocking them from what they want to accomplish. With this diagnosis, the problem appears to be one of devising ways to escape from the clutches of the bureaucracy or to overpower the bureaucracy by arguing for one's program.

Social scientists could help here because many of them have had some theoretical training in organizational behavior, an interdisciplinary field in which sociologists, social psychologists, anthropologists, and political scientists have been active. Unfortunately, the contributions of social scientists to the understanding of agricultural bureaucracies have thus far been extremely limited in the academic literature and practically nonexistent in the field. Most studies of organizational behavior have been focused on industry and government organizations in the more developed nations, particularly in the United States. Furthermore, the studies concentrate primarily on internal relations—on the problems of cooperation and conflict among workers, supervisors, and higher level officials working full time within the organization. The researchers have generally given little attention to delivery of service outside of the organization or to any relations with the outside world. Studies of relations between the organization and its environment are being made, but they tend to focus on the environment, seeking to discover whether it is "stable" or "turbulent." Such global characterizations give us little help in understanding the day-to-day interplay between insiders and outsiders. Two fine exceptions to this generalization are the studies of farmer-extension interactions in India by Stanley J. Heginbotham (1975) and in Kenya by David K. Leonard (1977). They look at bureaucratic interactions and attitudes within the "black box" as well as at why face-to-face relations with farmers are conducted as they are.

We do not claim that social scientists have available a large storehouse of knowledge on organizational behavior that is directly and readily applicable to problems of reorienting agricultural bureaucracies. We do see potentially important contributions by social scientists in describing and analyzing a variety of regional and national programs designed to build more participatory organizational systems. As social scientists gain this knowledge and experience, they will be able to build better theories of organizational change in agriculture and provide more useful technical assistance to those designing the participatory organizations of the future.

On Building Interdisciplinary Research Capacities

It should be possible for social scientists to make major contributions to the design of organizations and to the planning of social processes that will facilitate the development of interdisciplinary agricultural research. Social scientists have made such contributions to the planning of structures and social processes in the industrial sector. In this early stage of studies of agricultural bureaucracies, however, the most we can claim is that we have identified some of the major questions and are beginning to develop some tentative and very preliminary answers.

If we are to progress in this field, it is important that we clarify our objectives and recognize the difficulty of the problems we face. Regarding objectives, we should emphasize that we are not seeking to eliminate discipline-based research and to substitute interdisciplinary research. Because we have recognized the limitations of monocultural and discipline-based research, we have emphasized the importance of developing interdisciplinary research. But, the question is not whether one style of research is better than the other but rather one of how the two modes may be more effectively integrated.

The development of new plant varieties clearly calls for monocultural projects with highly trained specialists and with interdisciplinary contributions largely limited to various specialists in the plant and soil sciences. In laboratory and experiment station, highly trained specialists should continue to play leading roles in establishing research priorities based upon the assessment of the current state of knowledge in their particular disciplines. But, if project priorities on the experiment stations are established without any consideration of the problems and needs of small farmers, then past experience suggests that the scientific knowledge gained will bring limited benefit to the persons for whom it is intended and the payoff of research will be reduced.

A primary objective of farming systems research is to feed back information to experiment station scientists on the need for component technology. The follow-up may well be undertaken by specialists carrying out monocultural experiments, but they should be aiming to produce technology designed to fit into a farming system rather than attempting simply to produce a variety that is superior to others when grown in monocultural plots under experiment station conditions.

Highly trained specialists based in laboratories or experiment stations can also be essential resources for consultation on farming systems research in farmers' fields. No interdisciplinary team can possibly include all the specialists that might be needed to deal with any problem that might arise within a farming system. The specialist in the particular problem the interdisciplinary team is unable to solve does not need to be a member of the team in order to play a vital role as consultant.

Regarding the difficulty of introducing change, we should recognize that interdisciplinary problem-oriented research has not been widely practiced in the United States and is in a very formative stage in the developing world. The way universities and government research agencies are structured in the United States as well as in developing countries tends to reward individuals financially and through professional recognition of their perceived contributions to their particular discipline. Much time and effort and the development of new patterns of rewards will be required before it will be possible to build much capacity to establish effective interdisciplinary programs in developing countries.

19

Implications for Education

The reorientation of agricultural research and development programs to provide more effectively for the needs of small farmers requires modifications in the education and training of those involved in the work—professionals, technicians (paraprofessionals), farmers, and rural community participants. Since different levels of education, different training, and different institutions are concerned with the three groups, this chapter will consider them separately.

Professionals

The importance of reorientation of the educational programs of international centers, universities, agricultural colleges, and national ministries and departments concerned with agricultural development in the LDCs has been the concern of three workshops sponsored by the Rockefeller Foundation and the International Agricultural Development Service held in 1974, 1975, and 1979 at Bellagio, Italy (Rockefeller Foundation, 1974, 1975; International Agricultural Development Service [IADS], 1979). The discussions were centered on the needs for training of three categories of professional personnel: existing staff of national programs; mid-career senior scientific personnel concerned with planning and administration of national programs; and young national and foreign professionals preparing for service in national programs. The first and third categories have the same educational needs. The second, in addition to those in agricultural development, also needs improved competence in policy development, planning, and administration.

The basic capabilities needed by professional personnel (IADS, 1979) include understanding of tropical agriculture; orientation to problem-solving research; sense of urgency and awareness of problems; ability to handle farming skills; ability to conduct field experiments; understanding of the socioeconomic situations of the developing countries and the special needs of resource-poor farmers; ability to adapt to life in different cultures; ability to work as part of an interdisciplinary team; ability to forge linkages with institutions responsible for technology transfer; understanding of the compelling need to verify laboratory and experiment station results within a prevailing farming system in on-the-farm situations; ability to articulate; and insight into rural development strategy.

The institutional programs and opportunities available for agricultural development training include formal study for advanced degrees (M.S., Ph.D.) in national and foreign universities; special short-term training programs conducted by universities and colleges, international centers, and national ministries and agencies; and in-service training opportunities offered by the development programs or institutions and agencies that are involved.

Formal Study

The formal degree programs of national and foreign universities are available to qualified professional staff members and aspiring professionals who are preparing for service in development programs of their own countries or of international institutions concerned with agricultural development in developing countries.

Specialized education in the disciplines on which agricultural development depends is available in the more developed countries. The existing programs have provided the advanced training for most professionals preparing for agricultural work in LDCs during the post–World War II period. With material and technical help from the more developed countries, agricultural universities and colleges in some LDCs have developed professional training programs to the point that they can assume greater responsibility for education at this level. Notable among these countries are India, the Philippines, Mexico, and Brazil.

In the meantime, there has been increasing disillusion about the relevance of much of the training in developed countries for the needs of the LDCs. A recent book by Sterling Wortman and Ralph W. Cummings, Jr., gives an interpretation of the underlying causes for the disillusionment:

At least in the U.S. the universities face a dilemma. On the one hand, U.S. universities recognize the serious needs of developing countries, they recognize that their staff members can help alleviate food problems and can benefit professionally from experiences abroad and they recognize that some lessons abroad may have relevance at home. On the other hand, their primary responsibilities, the duties for which they receive public support, are to serve their own states, so overseas work is hard to justify. Their agricultural courses are becoming more narrowly specialized to serve the advanced agricultural systems of the U.S. Consequently, universities have fewer agricultural strategists and they are producing few of the broadly trained agronomists or animal production specialists developing countries need.   [1978:417]

The result of this dilemma is expressed by the Asian Development Bank:

The western educational system, which serves as the model for most is highly discipline-oriented. Researchers are rewarded for achievements within the narrow context of their discipline. The result is not only a lack of pragmatism in research, but also an extraordinary lack of communication among disciplines. The lack of communication and understanding is reflected in the distrust between biologists and social scientists, and even within these broad categories, the distrust among the disciplines. The lack of communication exists not only horizontally among disciplines, but also vertically among research workers, extension specialists and farmers.   [Cited in Wortman and Cummings, 1978:387]

Thus two general problems affecting the relevance of the conventional advanced-degree programs in universities of developed countries have been recognized: high specialization and accompanying lack of a broad orientation needed in development programs for LDCs, and absence of field experience on the LDC agricultural situation. The first of these—overspecialization—involves the program of studies at the university. The second required inclusion of LDC field experience as part of the training program.

The recognition of overspecialization as a problem has led to the creation of M.S. or Ph.D. programs in international studies in some American universities that provide special seminars and courses in which the students from such diverse fields as agriculture and biology, sociology and anthropology, economics and political science are given the opportunity to interact with each other and the staff in the study of specific rural agricultural development problems in less developed countries. At Cornell University, for instance, there are now seminars sponsored by the International Agriculture Program and the Rural Development Committee, in addition to the regular seminars in the departments. Interdisciplinary courses are presented on organization, administration, and management of development projects, and an annual study trip to Mexico provides opportunities to visit CIMMYT and some of the ongoing work of national institutions with farmers.

With respect to field experience, universities of developed countries are making arrangements for some of their students to do thesis work in less developed countries. In this case, cooperation of staff members and use of facilities of national institutions may be obtained through ad-hoc arrangements between the thesis advisers and their colleagues in the receiving institutions. The special funding needed for such programs now may be provided by foundations and international agencies. The Ford Foundation sponsored a Ph.D. program for Cornell graduate students, which provided for thesis work and some formal study at Los Baños, the Philippines, with the cooperation of IRRI and the college of agriculture in that country (Turk, 1980). More recently, CIMMYT sponsored field thesis work in Mexico by Cornell graduate students in international agriculture (Contreras et al., 1977). The 1979 Bellagio conference report states that universities of the United States, the United Kingdom, Netherlands, and France include some period of orientation in a developing country (IADS, 1979). Some seventeen countries contribute to the FAO's International Associate Expert Program, through which many young graduates receive on-the-job training in developing countries. All the international research centers, research internships for national program personnel, thesis research grants, and postdoctorate programs provide training opportunities in research relevant to the development and beneficial reciprocal linkages at student and staff levels. The 1979 Bellagio conference (IADS, 1979) noted, however, that inadequate numbers are being trained to conduct relevant agricultural research in developing areas; that the training staffs and facilities frequently are insufficiently oriented toward area development and on-farm research on the social, economic, and resource availability conditions of development in tropical areas; and that the team approach, group thinking, planning skills, and other factors relating to agricultural development are seldom included in training experiences for young professionals. Thus even though progress has been made in reorientation of educational programs of universities and international centers designed to serve agricultural development programs in the LDCs, this analysis stresses the need for "more and better." It is reassuring to find in a recent symposium held at Kansas State University in 1981 on the topic "Small Farms in a Changing World, Prospects for the 80's" that relevant technology development and dissemination at the international agricultural research centers (McDowell, 1981) continue to have high priority.

Special Short-term Training Programs for Professionals

The preparation and orientation of mid-career professionals for the planning and management of agricultural development programs in

LDCs have received less attention than the advanced training of young professionals. In the less developed countries, those who are chosen for these key responsibilities may have limited experience in agricultural research and technology planning or management procedures and skills. The expatriate professionals chosen for participation in agricultural development programs may have little understanding and experience, beyond their specialties, in the planning and management strategies and the linkages and team requirements of successful agricultural development programs for small farmers in LDCs.

Training programs for mid-career professionals have to conform to the time that these scientists and administrators can be away from their responsibilities. Senior government officials seldom can spare more than a week to ten days. The longest study leave for the others is probably about three months. Programs providing intensive study of agricultural research management, within these limits, are offered by a number of international and national institutions, both in more developed and less developed countries. At the time of the 1979 Bellagio workshop, they included programs of the Southeast Asian Regional Center for Graduate Study and Research in Agriculture, international agricultural centers (CIMMYT and IRRI), U.S. Department of Agriculture, World Bank, Harvard University International Courses for Development-Oriented Research in Agriculture (Europe) and, in two Indian institutions, programs concered with program management. These and other training programs for mid-career persons are still in their formative stage, but they are an encouraging indication of a widespread appreciation of the need for regional approaches to accelerated training for research managers. The study group of the 1979 Bellagio conference concluded that it is important to establish regional centers—initially in Southeast Asia and Latin America—for this training and that it would be desirable for a single agency such as International Agricultural Development Service to formulate and administer the program.

Perhaps the most perplexing problems to be dealt with are the short time available to participants in the training courses and the constantly changing national agricultural programs and the short tenure of leaders involved in their administration. Given these limitations, the courses themselves have to be regarded as a way of initiating interaction among leaders and introducing them to the international and regional resources available for the development of their national programs. The regional centers, therefore, will have to provide strong continuing linkages with the leaders who visit them briefly for special training and to be able to maintain library and documentary information service on a wide range of development activities not covered in individual courses. But it is also evident that courses and information sources provided by regional and international centers can be made more effective if they are complemented by in-service and on-the-job training at the national level. Little has been done thus far to recognize the need for interdisciplinary seminars and workshops to provide middle-level managers of development programs interaction in the planning and execution of their activities related to the needs of small farmers in LDCs.

Paraprofessionals and Technologists

Paraprofessionals and technologists engaged in agricultural development may have received education to the high school level. Some of them are graduates of two-year post–high school agricultural institutions, which train them in farm management, agronomy, and other subjects related to practical agriculture. A large proportion of the latter group find employment in extension, research, and rural service agencies of ministries and private entities serving farmers. In the reorientation of agricultural research and development programs, they play a key role, not only in the conventional transfer of technology to farmers but also in the feedback from farmers—the recognition and interpretation of the needs of and constraints on the small farmers—as a basis for the planning and execution of research.

The effectiveness of the paraprofessional in this enlarged role will depend upon his understanding of the agricultural development process and his place on the interdisciplinary research team. It will involve training not only in production agronomy, soil management, and pest control but also in the elements of farm management and rural sociology and the procedures of research methodology. Both the curricula and the subject matter of the individual courses in these two-year training programs need revision and strengthening. Since they are dealing with highly motivated, competent young technologists, there is an important need to provide a bridge between their programs and the undergraduate programs of universities training agricultural professionals, so that qualified paraprofessionals may advance to professional rank in the normal course of their training experience.

At present, some intensive courses are offered by international centers for the training of selected paraprofessionals in the conduct of cooperative plant improvement research. The eight-month training program of CIMMYT in corn and wheat improvement, IRRI on rice improvement, and CIAT in bean and cassava improvement are cases in point. These programs give opportunities for participation in the key field operations over a crop cycle. IRRI and ICRISAT also give training in cropping systems research for their respective climatic zones (CGIAR, 1976).

Without doubt in-service training will continue to be the most important educational tool at the paraprofessional level for those concerned with agricultural research and development. The most effective in-service training is a continuous process, involving a regular schedule of training meetings which relate to the work to be done in the immediate future and providing for exchanges of information about recent field problems. An example is given by Daniel Benor and James J. Harrison (1977) in their description of the training and visit system of agricultural extension. Much remains to be done, however, in developing in-service training that stresses interdisciplinary team research approaches and fully incorporates farmer participation in the planning, execution, and evaluation of field experiments and on-farm trials.

Farmer and Rural Community Participants

The successful reorientation of agricultural research and development programs for small farmers requires active participation of the farmers and the members of the rural community who provide essential services to the programs. The educational level of most of this group ranges upward from less than six years of elementary school to high school. Functional illiteracy is common; rural isolation is the usual condition. With these social disadvantages, the difference in status and power between the professional and these individuals is so great as to make communication difficult. The fact that professionals have been accountable to their employing bureaucracies rather than to the farmers has increased the breach in communication. It is through local farmers' organizations that the most rapid progress in rural and agricultural development has been made in Asian developing countries (Uphoff and Esman, 1974). And it is through these organizations that much of the training of small farmers for effective participation in agricultural development programs must take place. In addition, enhancement of the role of paraprofessionals in communication to and from farmers has provided a necessary link with professionals.

The process of this education has to take place, for the most part, in the course of participation experiences, which can be supplemented by meetings to discuss specific subjects, by field trips, and by occasional two- or three-day schools. For Chapters 13 and 14 we gave examples of ways in which ICTA in Guatemala, PRODERO in Honduras, and FORUSA in Mexico have provided for in-service training of participants in development programs. The examples illustrate the importance of institutional and social structures within each of these countries in determining the appropriate details of the training process.

Multiobjective Planning for Educational Development

Planning for improved educational programs should not be carried out in isolation from the planning of activities in other closely related fields. Planning should be based upon a theoretical principle concerning the potential economies of multiobjective planning. The principle simply states that, when the same or similar human and material resources can be used to progress toward two or more socially valued objectives, this strategy will lead to a more favorable balance of benefits over costs than can be obtained when the project is designed to pursue a single objective.

The Caqueza project (Zandstra, Swanberg, and Zulberti, 1976) provides examples of the implementation of such a strategy. The project provided funds and technical assistance for university students to carry out research leading to their graduate theses on problems of interest to Caqueza and to ICA, the national agricultural research institute of Colombia. The project yielded the following benefits: It increased opportunities for students to do field research on problems of interest to themselves as well as to Caqueza and ICA. It enabled students to advance toward graduate degrees and to become fully qualified professionals, thus enriching the human resource base for the further development of Colombian agriculture. The project also strengthened relations between university professors officially responsible for student theses and Caqueza and ICA, thus contributing to the development of more effective collaboration between the university and government agriculture research and development agencies.

The Caqueza program involved graduate students, and there was no immediate application of the research although Caqueza considered the problems under study to be ultimately of practical significance. Let us now turn to a project in Peru that involves action research and included undergraduates as well as graduate students and professors and that ties professionals and students directly with peasant farmers in the conduct of field activities (Whyte, 1977).

The director of the Huasahuasi project was Ulises Moreno, who had been born and brought up in that highland village. He went through the principal Peruvian agricultural university at La Molina just outside of Lima, and then came to Cornell for a doctorate in plant physiology, writing his dissertation on Peruvian potatoes.

Ulises Moreno may have been the first student from any developing country to return to his peasant community to report on his doctoral dissertation. For this event, he prepared very conscientiously, rehearsing the way he would put the practical implications of his dissertation into language the villagers could understand. Moreno presented his talk one evening in the central square of the town, projecting slides that showed the various aspects of potato physiology on a scale the villagers had never seen before. Afterward, many villagers gathered around Moreno to urge him to organize a project to help the small farmers of Huasahuasi.

When Moreno took up his professorship at La Molina (Universidad Agraria Nacional), the university had a contractual relationship with the Ministry of Agriculture to support technical assistance to the communities. In 1973, Moreno secured a grant of $2,000 to be used primarily for the travel expenses of his La Molina team to Huasahuasi. The villagers of Huasahuasi pledged room and board for the team while it was in that community. The National Potato Program made contributions of improved seed varieties. Moreno arranged for the participation of extension agents working in the region around Huasahuasi. A Bayer Chemical Company agent provided chemical materials and agreed to demonstrate their use.

The agricultural economy of Huasahuasi is based upon potatoes. The municipality produced more than half of the seed potatoes used by coastal potato farmers. All of these seed potatoes were grown by the big farmers, however, the largest one having up to 7,000 hectares. The big farmers were getting up to forty tons of potatoes per hectare, while the small ones (without improved seeds and other inputs) were getting about ten tons (which was nevertheless above the national average).

The detailed story of this project is told elsewhere (Whyte, 1977). Here we present a brief summary.

To launch the project Moreno organized a team consisting of five professors, four graduate students, and seven undergraduates. The professors represented plant physiology, plant pathology, entomology, agricultural extension, and sociology; the students were majoring in agronomy, biology, and economics and planning. After a week devoted to orientation and planning at La Molina in August 1973, the team spent two weeks in and around Hausahuasi. Throughout this period, team members walked the fields, observing and interviewing farmers regarding their experiences and problems in growing potatoes. The individual efforts were followed up by small group meetings in each neighborhood of the municipality. Technical aspects of potato cultivation formed the central focus, but villagers also spoke about their needs for transportation to markets, credit, improved seed varieties, and more effective cooperation among themselves. They further discussed the quality and quantity of services they were receiving from the government.

The second team visit to Huasahuasi took place in February 1974, during the university vacation period. This time was devoted to joint planning for an experimental project in which local community government would assume substantial responsibility, with the technical assistance of the team from La Molina.

The team returned to Huasahuasi in October to carry out the planting of the experimental plots. The planting was accompanied by extensive discussion with the peasants regarding the logic of the experiments: what was to be done in relation to what they hoped to find out. All of the team members shared fully in the manual labor involved in preparing the fields and planting the potatoes.

The planting was done according to an experimental design permitting later comparison of improved seed varieties with the locally popular variety, along with variation in the use of fertilizer, the impact of fertilizer with and without insecticides, and a comparison of results when rows were spaced 40 cm or 30 cm apart.

The team returned again several months later to supervise and participate in the harvest. The work was done so that, in each row, as the potatoes were dug, they were laid out for all to observe. Then the villagers and the university team weighed and recorded the yield of each experimental plot for comparison purposes.

Finally, the potatoes harvested were divided among villagers who had volunteered their labor at every stage of the experiment including the harvest. For the first time, participating peasants had their own supply of seed potatoes of the most high-yielding varieties. Even before weighing the yields, the villagers could see the major improvements achieved with improved seed potatoes and optimal application of fertilizer and insecticides. Moreno later calculated that the yield of the improved varieties under optimal conditions indicated a potential yield of forty tons per hectare—equal to the best record of large farmers in that area.

Previously the large producers of seed potatoes had been unwilling to provide seed potatoes to the small farmers of Huasahuasi even when they hired small farmers in the planting or harvesting of their fields. They had a good market for those seed potatoes on the coast, and they recognized that, if small farmers got enough of the best variety and credit to buy fertilizer and insecticides so that they could quadruple their yields, they would need to devote more labor to their own fields and would have less time and interest in working for the big operators. Also, up to this time, the agricultural bank had not considered crop loans to small farmers as a practical use of money because the local prejudices indicated that small farmers were a poor risk. Now some of the small farmers had the seed potatoes they needed, and the widely publicized results of the experiment, along with the personal and institutional connections between the university and government, could be used to overcome the bankers' resistance to providing credit for the small farmers of Huasahuasi.

How do we evaluate this project? We got opposite answers according to whether we assess it in terms of a single objective or of multiple objectives. For example, one critic commented: "So Ulises Moreno has proved that with technical assistance and good seeds and fertilizers and insecticides, the small farmers can greatly increase their yields. What is new about that?"

The problem here is that the critic mislabeled the project, evaluating Huasahuasi as an experiment in the plant sciences. Clearly, the project could not tell the plant scientist anything about the technical aspects of growing potatoes that he did not already know. The picture changes when we view the project as a social experiment: the development of a new organizational model for intervention in peasant agriculture.

Furthermore, the critic was evaluating the project in terms of one objective, whereas this was a multiobjective project. To evaluate such a project, we have to consider at least the following seven objectives:

(1) Help the small farmers of Huasahuasi to do as well in potato growing as the large farmers.

(2) Develop and test a new organizational model for research and development in agriculture.

(3) Discover the social and economic barriers to progress in a community such as Huasahuasi, thus combining social knowledge with knowledge of plant sciences.

(4) Improve the education of students and professors. Perhaps the most important lesson learned involved humility. One young woman said to Moreno: "This has been a fascinating but also a frustrating experience. I have learned so much from the farmers of Huasahuasi, and at the same time I have had to recognize that for all of my years of study in biology, plant pathology, calculus, and so on, there is very little that I have been able to contribute to them." If students can learn this lesson before it is too late, they should be able to contribute far more than those whose university education has been limited to classroom and laboratory.

(5) Stimulate interdisciplinary communication for both professors

and students. Research and development in agriculture require the skills and knowledge of specialists from various disciplines. University programs tend to develop specialization; rarely do professors from one field work with those in a different specialty. Moreno has demonstrated one way of stimulating interdisciplinary communication and collaboration.

(6) Improve relations among the university, the national and international research programs, and agricultural extension. In Peru as in many other developing countries, the activities of these various institutions tend to proceed more or less in isolation from each other.

(7) Improve the effectiveness of agricultural extension. As extension agents participate in projects such as Moreno's, they should learn how to make a more substantial contribution to agricultural development.

Although the benefits to be gained from the second through seventh objectives are difficult to quantify, they nevertheless are built to meet deficiencies and needs that have been widely recognized in research, development, and education in many countries.

The University as Focal Point for Change

The changes we have called for can be initiated in various parts of the system of agriculture-related institutions, but we think that the university in the developing country as well as in the developed country may be a strategic point from which to launch the innovation process. To be sure, universities in developing countries face serious handicaps. They are often so poorly financed as to be unable to provide professors and students with adequate laboratory and library facilities. Professional salaries may be so low as to force many or even most of the instructional staff to work at other jobs in order to attain a middle-class standard of living.

On the other hand, more than other institutions, the university may have the flexibility to stimulate creativity and innovation. The university is dedicated to scientific progress, which means to professors and students that today's knowledge will be outdated tomorrow. Many professors will continue to teach outmoded ideas from outdated textbooks, but there will always be some who will be troubled by the university's lag behind advancing knowledge and eager to change their methods of teaching and research.

Compared to previous generations, today's students are less likely to accept academic traditions unquestioningly. Increasing numbers are demanding that their education become "relevant" to the serious socioeconomic problems of the nations. Even if many of those students interpret relevance as being library study of theories of revolution, there will always be some who, whatever their degree of commitment to drastic social change, will be willing to see relevance in gaining firsthand field experience with small farmers. As in the case of Ulises Moreno, if a professor is willing to lead them into the field, they will follow with enthusiasm. In universities of both developed and developing countries, disciplinary specialization will continue to be a barrier against the interdisciplinary teamwork required in farming systems research and development. On the other hand, it is much more efficient from a societal standpoint to seek to break the pattern of excessive specialization in higher education than to try to resocialize the agricultural scientist for interdisciplinary teamwork after he or she begins to work in a national or international research organization. Of course, these organizations continue to consider first whether the prospective employee is competent in some specialty, but they are increasingly looking for students who have had some interdisciplinary experience and complaining that many people they hire are too narrowly trained and too inflexible to be effective in working with specialists in other disciplines. We hear leaders in such organizations complaining that they have to supply the interdisciplinary orientation that students generally fail to get in the universities. Thus we can expect increasing pressures from such research organizations upon the universities to foster opportunities for interdisciplinary learning.

Although government bureaucracies may be so rigidly structured that any major change is impossible without strong and continuing support from the top, the university provides for more autonomy and freedom of movement to deans, department heads, and even individual professors. As in the case of the Huasahuasi project, an enterprising professor may inspire a few other professors and students to join in an innovative research and development project. Even if the innovation is on a small scale, it may attract enough attention among university administrators, professors, and students to stimulate serious discussion of how the university can plan and support an expanded program growing out of what has been learned in the initial project. In fact, the Huasahuasi project became a major element in planning at La Molina for the development of a program whereby all students would gain substantial field experience in rural villages before getting their undergraduate degrees.

Since money is scarce in most developing countries, it is important to recognize that, by combining education with a field project, the university can often accomplish more with less money than can a government agency. The total cost to La Molina of the Huasahuasi team was $2,000. If a government agency had provided the same size team for three periods of intensive field work in the village, the cost would have been many times greater. The major difference in cost is because the government agency would have had to pay salaries to its field team members whereas the students and professors volunteered their time. In working without pay, students will not feel exploited if they believe that the field experience is substantially enriching their education. Even if they are sacrificing opportunities to supplement their salaries in part-time jobs in the city, professors will not feel exploited if they believe they are participating in an important innovation—and also if they think that participation will win them recognition from colleagues and university administrators.

As in the La Molina case, a single professor may take the initiative in educational innovation. Unless the project then wins the approval and support of department heads and university administration, however, it will remain simply an isolated case rather than the beginning of a broad innovative program.

Although the strategy of multiobjective planning may be attempted in any organization, we see the university as an especially fruitful setting. The relatively greater flexibility of the university makes it easier for a professor or administrator to devise such a strategy and then reach out to other university units and to agricultural research and extension agencies for collaboration and financial support. Therefore, we can expect some universities to be in the forefront of change, not only devising new and better ways to teach but also creating new models to combine teaching with research and with development projects.

Conclusions

This chapter has dealt with the modifications in the education and training of professionals, technicians, farmers, and local leaders that are needed to provide for more effective small farmer research and development activities. The objective is to foster better horizontal coordination between the physical, biological, and socioeconomic approaches and better vertical coordination between the planners, researchers, extension, and farmers involved in the development work. Training in interdisciplinary teamwork and organization to foster farmer participation and feedback are essential parts of this educational process.

For young professionals considerable progress has been made in the offering of university programs that provide for interdisciplinary approaches to problems of farmers in developing countries. Not only are formal course programs available at the undergraduate and graduate levels in an increasing number of universities in developed countries, but field experiences and some course work are available in some less developed countries.

For mid-career professionals, there is an increasing number of opportunities for participation by planners, administrators, and executives of agricultural development organizations in interdisciplinary seminars, roundtables, and short courses concerned with aspects of their work. The international agricultural centers and several university centers serve as important information resources for training and exchange of experience at the mid-professional level.

Technical training at the paraprofessional level in international agricultural centers and in national institutions of less developed countries provides increasing opportunity for this important group to serve as a link between farmers and research workers in the feedback of essential information from the small farmer to the agricultural planner and research worker.

Training of small farmers and rural leaders for their key role as participants in agricultural development research has to occur largely at the local level. Farmer cooperative associations and other rural organizations largely form the basis for this in-service training.

Although real progress can be seen at all of these levels, it is evident that the last link in the chain—at the farmer level—needs the most strengthening. Because of their social and educational handicaps, as well as their rural isolation, small farmers need to be given far greater educational opportunities at the same time that their essential role in agricultural development is being recognized. Because the small farmer is likely to be closest to the paraprofessional working on agricultural development, it is at this level that much additional educational opportunity is required. At the top of the chain—the young and mid-career professionals—the crucial need is for a reorientation of the training process so as to provide for more effective interdisciplinary activity and better linkages with small farmers through the medium of paraprofessionals.

20

Implications for Government Policy

We have ranged widely around the world and across many disciplines, but always we have sought to focus on problems, processes, and potentialities for active involvement of small farmers in the agricultural research and development process. Finally, we seek to draw general lessons for government policy from what we have learned.

A Central Role for Agriculture in Research and Development Policies

The first requirement for building the kind of agricultural research and development program which we have been visualizing is a strong government commitment to the central role of agriculture in its research and development policies. This need may seem obvious now, but, even though no government ever neglected agriculture in its political rhetoric, in fact, at least up through the 1960s in most developing countries, agriculture was a stepchild among development planners. They equated modernization with industrialization. Rural areas were expected to supply labor, food, capital, and taxes to the "modern" sector; as our colleague Erik Thorbecke puts it (personal communication), agriculture was regarded as a goose to lay the golden egg for industrial development.

In this era, development planners regarded rural areas as overpopulated by people who farmed inefficiently on small holdings. The idea was that the growth of industry would provide jobs for the surplus rural population, and migration away from the rural areas would lead to consolidation of agricultural land into larger and therefore more efficient farming units.

This idea has since been discredited, in part by the course of events and in part by research. Urban migration has indeed continued, but in few, if any, countries has it resulted in a decrease in the rural population. Nor has the growth of industry been rapid enough to provide jobs for the migrants. Furthermore, all too often planners of industrial de-

310

velopment, lured by the example of industrialized nations, have pushed capital-intensive "modern" technology that provides few jobs and have neglected to exploit the potential of intersectoral linkages between agriculture and industry through promoting industries using agricultural raw materials or producing inputs for agriculture. Finally, research in various countries has destroyed the old stereotype of the inefficient small farm. Researchers are finding that small farmers frequently either achieve higher yields per hectare than large farmers, or, at least, make more efficient use of their limited inputs than do larger farmers.

The skyrocketing prices of oil in the decade of the 1970s have also been a powerful influence in redirecting the attention of government planners back to agriculture. For those developing nations without substantial oil resources, petroleum imports become an enormous drain on foreign exchange. Pressed to the limit for funds to import fuel and industrial products, developing nations can ill afford to import food. Under these conditions, self-sufficiency in agricultural production becomes a primary development goal.

Dominance of the Socioeconomic Structure

The socioeconomic structure of the nation tends to shape the distribution of both political power and material resources. Unless strong measures are taken to overcome these structural biases, the rich tend to get richer and the poor tend to be largely left out of the development process. Many critics have seen land reform as a key to basic socioeconomic change. Without minimizing the importance of land reform—a topic largely beyond the scope of this book—we argue that changes in land tenure need to be seen in a broader socioeconomic framework. We must also consider the distribution of other resources and of information and ideas about using those resources. We have stressed the importance of the organization of small farmers and the rural poor in general to provide a political and economic base to improve the position of lower-class rural people in the socioeconomic structure of the nation. We consider the elements that must go with land reform in subsequent sections of this chapter.

Research and Development Strategy for the Active Involvement of Small Farmers

Policy makers are coming to recognize that large farmers, who have received the lion's share of government resources and attention in the past, can get along well enough without being so favored in the future.

Furthermore, the most spectacular gains of the Green Revolution have been achieved on large irrigated lands, and the potential for improving productivity on these lands may now be limited. For example, in Mexico, the large wheat farmers on the irrigated lands of the northwest increased their yields so rapidly in the 1960s as to make the nation self-sufficient in wheat. Apparently, those farmers continue to be highly efficient, but they have not been able to increase yields to keep up with the growth of population, so that Mexico has once again become an importer of wheat. In maize, the food deficit has grown so large as to alarm government policy makers and produce a major emphasis upon increases in yields. Most of Mexico's maize is grown on small farms, so a strategy based upon the development of this sector is required.

Having been forced to give special attention to small farmers, Mexican planners are now much more optimistic about the prospects for improvement in that sector than were their counterparts in earlier decades, when the prevailing view was that small farmers were tradition-bound and almost impervious to change.

Policy makers in agricultural research and development are coming to recognize the efficiency and productivity of the small farm. They are abandoning the view that the small farmer is an irrational prisoner of his backward culture. They are recognizing that there is a compelling logic to the strategies and practices of the farm family. They are recognizing that the traditional monocultural emphasis in agricultural research does not fit the needs of the small farmer who is developing a complex pattern of crops and animal husbandry. This new emphasis upon cropping and farming systems research can be seen in a growing number of countries.

Principles of the New Research and Development Strategy

The first principle of this emerging research and development strategy has been stated by Randolph Barker: "the baseline is not zero; the baseline is the farming system currently being practiced by the small farmer" (personal communication). The first step therefore calls for concentrated study of indigenous farming systems. It is only on the basis of a solid understanding of indigenous farming systems that researchers can efficiently contribute to improvement.

In contrast to past research with its monocultural production emphasis, the emerging R&D strategy is based upon multiobjective planning. Depending upon local needs and the interests of participating farmers, the agricultural professionals may focus on increasing yields of particular crops, on multicropping combinations, on more efficient integration of animal husbandry with the cropping system, on improvements in storage, marketing, and credit, and on developing better methods for land and water conservation. In other words, the professional does not simply move in on the community to promote his own specialty. He develops plans only in the process of assessing local needs and interests, and he carries out this assessment in collaboration with local people who feel those needs and have those interests. Broad as this range of needs and interests may be, the agricultural professional must also recognize the potential importance of other projects that are only indirectly related to agriculture.

This does not mean that the agricultural professional must concentrate on rural electrification, on improvements in public health, or on a literacy campaign, or whatever happens to be the number one priority of the village. It does mean that, while continuing to work in the general area of agricultural research and development, a professional recognizes that the villagers may have other priorities and, respecting their right to make their own judgment, he at least seeks to facilitate their contacts with individuals and agencies who may help them meet those needs. Of course, it will be exceedingly difficult for the agricultural professional to help the villagers make connections in these other areas unless the government is involved in an integrated program of rural development.

The new strategy is based upon a critical shift in terminology and ideas that may be expressed as moving away from technology transfer and toward the facilitation of progressive change. The term *technology transfer* implies that the professional enters the scene equipped with technologies of proven value and needs only to apply these technologies on the local scene, while making minor adaptations as he goes along. This formulation also is implicitly based upon the idea that the professional must tell the small farmers what to do and somehow persuade them to do it.

In contrast, the new strategy is based upon the assumption that small farmers know what they are doing but are willing to consider new options. In this context the aim of the professionals should not be to make choices for the small farmers but rather to broaden the range of available choices. This strategy implies deemphasizing demonstration projects, which are based on showing farmers what they should do in the use of plants and animals, farming practices, and inputs. When demonstrations are used, they should be put on in collaboration with small farmers and as a means of evaluating a new option rather than for the purpose of telling the farmers that this is what they must do.

Except in rare cases (for example, when the addition of a trace element previously lacking in the soil makes a dramatic improvement in yields), improvements in agriculture are not achieved simply by the addition of any single input or the change of any single practice. Agricultural research and development depends upon understanding the complex interrelations of the variables with which the farmer actually deals. Recognizing these complexities, some advocate presenting the farmers with packages of inputs and practices. Whether this constitutes an improvement depends on how the package approach is interpreted. Agricultural professionals are increasingly recognizing the extraordinary variability of conditions of soil, water, and climate so that a package that proves rewarding in one area may be ineffective even in a neighboring area. Therefore, the objectives of government policy should be to help the farmer work out, in collaboration with professionals, the combination that best meets the ecological conditions under which he farms and the socioeconomic conditions and interests of his family.

Linking Research and Extension

Implementing the new strategy requires breaking down traditional barriers based upon a one-way flow of information and ideas from researchers to extension agents. As we have seen in Honduras, extension agents can become involved in the study of indigenous farming systems. They can participate with researchers and farmers in planning and implementing on-farm experiments. They can also help small farmers to organize so as to strengthen the resource base that supports them economically and enhances their influence politically. Government policies must enhance the role of the extension agent as a communications link between researchers and farmers both in technology transfer and in feedback.

Utilization of Paraprofessionals and Farmer Leaders

We have noted the enormous possibilities of strengthening agricultural research and development programs through the increased utilization of paraprofessionals and farmer leaders. To be sure, how to use these new categories of personnel is a problem that practitioners and researchers are still trying to work out, but the economic potential is clearly great. For example, we sometimes find policy makers in the capital debating among themselves as to how much of a premium they have to offer professionals to persuade them to work in the "boondocks." Fortunately, paraprofessionals and farmer leaders do not know that they are in the boondocks and do not have to be given special inducements to remain where they want to be. Therefore, policy planners will find that, through a more efficient and widespread use of paraprofessionals and farmer leaders, they can multiply the impact of their program without multiplying costs.

Building Organizational Linkages

Agricultural development cannot be carried out by agricultural professionals alone, even if they work effectively with small farmers. The development of any area will depend in part upon organizational linkages connecting research, extension, credit, education, public works, and so on. This principle is generally well recognized, but all too often in the past top government policy makers have sought to solve the problem by establishing a high-level coordinating committee in the capital. As we have seen, top-level coordination may be necessary, but it is certainly not sufficient to clear the obstacles of interorganizational conflict. Coordination must be developed especially at regional and area levels. Furthermore, such coordination cannot be established solely by designing an organization structure and providing for coordinating committees. Leaders must develop a management process in which representatives of each agency having a role in agricultural and rural development meet regularly to plan, to report on progress and problems, and to work out the formal and informal arrangements necessary for interorganizational collaboration.

Organizational Decentralization

If we emphasize developing programs to benefit small farmers, we necessarily go beyond the traditional abstract arguments in favor of decentralization. Given the enormous variability in soil, water, and general ecological and socioeconomic conditions, no centralized program of agricultural research and development can meet the needs of the countryside. A program based upon responsiveness to the needs and interests of small farmers must necessarily be decentralized, with the professionals, paraprofessionals, and farmer leaders who work directly with small farmers having considerable autonomy in decision making. It will be extremely difficult to design and implement a decentralized agricultural R&D program when the government itself is highly centralized. Here we are not only drawing attention to the problems that a decentralized ministry of agriculture would have in coordinating its activities with other, more centralized government bureaucracies. Unless local and regional government officials have some degree of autonomy and some resources under their own control, it will be exceedingly difficult to secure the benefits of decentralized and responsive agricultural administration on the countryside.

Small Farmer Organization and Participation

When small farmers simply are recipients of initiatives from those above them in the socioeconomic scale, we cannot expect them to gain the benefits that come through active participation in which they play major roles in helping themselves.

In examining the case of ICTA in Guatemala we have noted that it is possible for an agricultural research organization to develop on the farmers' fields a process of experimentation in which small farmers participate actively. Such participation, however, depends upon the voluntary decisions of government officials to facilitate it. It is built on a fragile base and can be destroyed when key officials who believe in it are replaced by others who do not see the value of small farmer participation or even feel threatened by it.

The forces supporting participation are much more solidly based when small farmers come together to build a strong organization. In that case, if an official who believes in peasant participation is succeeded by one who is unsympathetic, the small farmers will not stand by passively. They will put pressure on the new official to respond to their interests and needs, and, if he fails to respond, they will seek ways to go above him and put pressure on higher-level officials.

When speaking of the ability of farmers to influence government officials, we are thinking politically. Farmer organizations also have important economic significance. Here we are thinking of economies of scale. When the small farmers are unorganized, the large farmer has great advantages over them in buying inputs and selling produce. He can get to markets inaccessible to the small farmer, and, since he buys in large volume, he has a much better chance than the individual small farmer to get the particular chemical composition of fertilizer most appropriate for his farm—quite apart from getting a discount for quantity purchases. Also, he can do his own marketing and get better prices than the small farmer gets dealing through intermediaries.

To a considerable extent, small farmers can overcome these disadvantages through organization. An efficient organization may bring them the economies of scale previously open only to the large farmer. "Efficient" is the key word, however, for it is certainly true that more small farmers cooperative organizations fail than succeed. Unfortunately, researchers still know relatively little about the factors that make the difference between successful and unsuccessful cooperative organizations.

As we have pointed out, the larger farmer also has important advantages over unorganized small farmers in access to information and technical assistance. Since the one-on-one extension-farmer relationship is inherently expensive in terms of the number of people contacted by the agent, there is a natural tendency for the agent to extend the geographical scope of this work through concentrating attention especially upon large farmers. Organized small farmers can counteract this tendency.

We know enough now to suggest two keys to the strength of the base-level local organization: control over its own resources and broadening of the decision-making options for members and for the organization.

Increasing local organizational control over resources weakens the dominance/dependency relationships that have traditionally prevailed between the economic and political elite and the peasants in developing countries. A simple illustration of this point appears in the Honduras case in which leaders of PRODERO distributed supplies of seeds under a plan that enabled farmers' associations to finance a major part of the agricultural needs of their members and thus liberate them from dependence upon the agricultural bank.

The participation emphasis also involves a marked shift away from the one-way initiative, in which professionals tell small farmers what to do and try to persuade them to do it, toward a situation in which the role of the professional is to help the villagers to discover and evaluate two or more options that might be to their benefit. If small farmers can get ahead only through following the instructions of the professional, this relationship reinforces their dependency. As small farmers gain experience in considering a broad range of options, they not only liberate themselves from extreme dependence upon professionals but also gain greater self-confidence, which enhances their abilities to help themselves.

From Pilot Project to National Program

When government leaders are persuaded that a small pilot project has achieved success, they naturally wish to expand that project as fast as possible to a regional or even to a national scale. The literature is full of failure cases in which expansion was carried out so rapidly that activities at the regional or national level came to bear little resemblance to what had been done at the local level in the pilot project. Any innovative pilot project involves a good deal of learning through experience by the original project planners and implementers. Furthermore, those who have invented the project naturally feel more commitment to the idea than do those later hired simply to follow the guidelines. Therefore, those involved in extending the pilot project must recognize the serious problems in recruitment and selection, in training the new professionals, and, above all, in developing a process of organizational change which will enable them to move slowly but steadily from the local level toward a broader impact program.

Political Support

Since the emerging new system of agricultural research and development is not as easy to dramatize as a new high-yielding plant variety, the building of a reservoir and irrigation project, and so on, it is likely to be much more difficult to gain political support than is the case for more traditional approaches. Furthermore, consistency of political support is far more important for developing and sustaining a new system of agricultural R&D than it is for continuing the traditional systems.

Interdisciplinary Research Emphasis

In the emerging new system of agricultural research and development, the emphasis is upon interdisciplinary research, with plant breeders, plant pathologists, entomologists, soil scientists, animal scientists, and agronomists working together to study cropping and farming systems. Social scientists also play important roles in linking the socioeconomic aspects of farming with the natural science aspects. Social scientists increasingly contribute not only at the community level in studying local family, community, and social structure but also study farmer organizations and government bureaucracies in relation to agricultural and rural development.

The new system does not eliminate specialization. There will still be need for a highly qualified specialist who can be called upon to diagnose problems of plant pathology, of insect infestations, of agricultural marketing, and so on. But those who wish to limit their work exclusively to their specialty will not serve effectively as members of teams studying and developing cropping and farming systems. They may be indispensable as consultants on particular problems, but the team needs to be made up of individuals who, however strong their specialized knowledge, are interested in interdisciplinary collaboration.

Linking Research with Education

Students will continue to need firm grounding in some particular discipline. Otherwise, they will not feel confident of their knowledge in any aspect of their work, nor will they be recognized as full-scale professionals by their colleagues. But if university students have no experience on an interdisciplinary project in the course of their education, any national or international program that emphasizes interdisciplinary teamwork must undertake the difficult task of reeducating them. Therefore, education should include firsthand field work experience in which students are members of interdisciplinary teams. Of course, in line with our emphasis on small farmer participation, it is also important that the field work involve students in situations that require them to learn from and to develop skills in working cooperatively with small farmers.

Conclusion

The points discussed above should not be seen as isolated ideas. They represent a pattern reflecting the thinking of those in the field who are blazing the trail toward a new participatory system of agricultural research and development. Further advances in this system will depend on understanding the pattern of thinking on which it is based.

The changes we recommend are of such a magnitude as to lead one to be pessimistic about the future rate of progress. On the other hand, we see two solid reasons for optimism.

In the first place, all over the world and in industry as well as agriculture, participation is an accepted idea. In a variety of forms, programs to stimulate worker participation in decision making in industry have come to be supported by legislation in Europe. In the United States, some of our major companies are emphasizing new worker participation programs so as to be better able to compete with the Japanese program of worker involvement in quality control circles.

In our foreign assistance program, the Congress has directed AID to give major emphasis to the active participation of poor people in projects designed for their benefit. In developing countries, government leaders are increasingly using the rhetoric of participation in their public pronouncements. To be sure, this is often empty rhetoric, with the political leaders having little commitment to the idea. Nevertheless, the idea is contagious, and we can expect rising levels of expectations in the countryside to be accompanied by increasingly insistent demands for active participation by small farmers.

In the second place, we have not been simply laying out our ideas of what ought to be done. We have seen in place major elements of this emerging new system of agricultural research and development in various countries in Asia, Africa, and Latin America. In case after case, we have seen researchers and small farmers overcoming the common confusion between education and intelligence. The traditional system was based upon the implicit equation of education with intelligence: the assumption that only those with a high level of education could be expected to have the intelligence necessary to lead small farmers toward a better life. As researchers have discovered the fine adjustments of peasant farming systems to their ecological conditions and have come to respect the resourcefulness and creativity of small farmers, they are demonstrating that the intelligence of small farmers must be the foundation of the human resources used to build better farming systems.

If this book makes a contribution to theory and practice, it will not be because we have presented new ideas. In large measure, the ideas we have laid out here we have found in the experience and practice of researchers and small farmers around the world. We saw an implicit consensus regarding the nature of the emerging new system across a wide range of disciplines from the plant, animal, and soil sciences to the social sciences. It has been our aim to pull together what such a wide range of people have been doing, thinking, and writing so as to articulate in a comprehensive and systematic form the nature of the emerging system of agricultural R&D necessary to enable the small farmers to share in the advances of science and technology.

References

Alfonso, Felipe B. 1981. "Assisting Farmer Controlled Development of Communal Irrigation Systems." In David C. Korten and Felipe B. Alfonso, eds., *Bureaucracy and the Poor: Closing the Gap.* Singapore: McGraw-Hill International.

Ali, Hashim Syed. 1980. "Water Management in Command Areas of Andra Pradesh, India." Paper presented at FAO Meeting on Farm Water Management, Beltsville, Maryland.

Anderson, Robert S., Paul R. Brass, Edwin Levy, and Barrie M. Morrison, eds. 1982. *Science, Politics, and the Agricultural Revolution in Asia.* Boulder, Colo.: Westview.

Ashby, Jacqueline. 1980. "Small Farms in Transition: Changes in Agriculture, Schooling and Employment in the Hills of Nepal." Ph.D. dissertation, Cornell University.

Asian Development Bank. 1977. *Asian Agricultural Survey, 1976.* Manila: Asian Development Bank.

Axinn, George. 1978. "Agricultural Research, Extension Services and Field Station." *International Encyclopedia of Higher Education,* pp. 241–254.

Axinn, George, and S. S. Thorat, 1972. *Modernization in World Agriculture: A Comparative Study of Agricultural Extension Systems.* New York: Praeger.

Bagadion, Benjamin J., and Frances F. Korten. 1979. "Government Assistance to Communal Irrigation in the Philippines: Facts, History and Current Issues." *Philippine Agricultural Engineering Journal* 10:5–9.

Barnett, M. L. 1980. *Livestock, Rice and Culture.* Bellagio Conference on Integrated Crop and Animal Production, 1978. New York: Rockefeller Foundation.

Beers, Harold W. 1971. *An American Experience in Indonesia: The University of Kentucky Experience with the Agricultural University at Bogor.* Lexington: University Press of Kentucky.

Benor, Daniel, and James G. Harrison. 1977. *Agricultural Extension: The Training and Visit System.* Washington, D.C.: World Bank.

Binswanger, Hans P. 1978. *The Economics of Tractors in South Asia.* New York: Agricultural Development Council.

Binswanger, Hans P., Vernon W. Ruttan, et al. 1978. *Induced Innovation.* Baltimore: Johns Hopkins University Press.

Black, John D. 1959. "Economics for Agriculture." In *Selected Writings of John D. Black,* ed. James P. Cavin. Cambridge, Mass.: Harvard University Press.

Blair, Harry W. 1974. *The Elusiveness of Equity: Institutional Approaches to Rural Development in Bangladesh.* Ithaca: Rural Development Committee, Cornell University.

_____. 1978. "Rural Development, Class Structure, and Bureaucracy in Bangladesh." *World Development* 6(1): 65–83.

_____. 1982. *The Political Economy of Participation in Local Development Programs: Short-Term Impasse and Long-Term Change in South Asia and the U.S. from the 1950s to the 1970s.* Ithaca: Rural Development Committee, Cornell University.

Boulding, Elise. 1977. *Women in the Twentieth Century World.* New York: Wiley.

Bradfield, Stillman. 1980. "Appropriate Methodology for Appropriate Technology." Paper presented at American Society of Agronomy, Chicago.

Brush, S. B. 1977. *Mountain, Field and Family: The Economy and Human Ecology of an Andean Valley.* Philadelphia: University of Pennsylvania Press.

Burchard, Roderick E. 1974. "Coca y trueque de alimentos." In Georgio Alberti and Enrique Meyer, eds., *Reciprocidad e intercambio en los Andes Peruanos.* Lima: Instituto de Estudios Peruanos.

Buringh, P. 1982. *Potentials of World Soils for Agricultural Production. Transactions of the 12th International Congress of Soil Science.* Plenary Session: 33–41. New Delhi, India.

Cary, Lee J. 1970. *Community Development as a Process.* Columbia: University of Missouri Press.

CATIE. 1979. "Small Farmers Cropping Systems for Central America." Final Report June 1975–March 1979, Contract no. ADI-596-153 (mimeo). Turrialba, Costa Rica: CATIE-ROCAP.

Chambers, Robert. 1974. *Managing Rural Development: Ideas and Experiences from East Africa.* Uppsala: Scandinavian Institute of African Studies.

_____. 1980. *Understanding Professionals, Small Farmers and Scientists.* International Agricultural Development Service Occasional Paper. New York: IADS.

_____, ed. 1979. *Institute for Development Studies Bulletin* 10(2).

Chancellor, William J. 1971. "Mechanization of Small Farms in Thailand and Malaysia by Tractor Hire Services." *Transactions of the American Society of Agricultural Engineers.* 14(5): 847–854, 859.

Chandler, Robert F., Jr. 1975. "Case History of IRRI's Research Management during the Period 1960 to 1972." Taiwan, China: Asian Vegetable Research and Development Center.

Cheng, Chien-pan. 1975. *Multiple Cropping Systems and Practices in Taiwan.* Sino-U.S. Joint Commission on Rural Reconstruction. Micro PID-C-566.

CIMMYT. 1974. *The Puebla Project: Seven Years of Experience, 1967–73.* Mexico, D. F., Mexico: CIMMYT.

_____. 1980. *Planning Technologies Appropriate to Farmers: Concepts and Procedures.* Mexico, D.F., Mexico: CIMMYT.

Cohen, John M. 1975. "Effects of Green Revolution Strategies on Tenants and Small-Scale Landowners in the Chilalo Region of Ethiopia." *Journal of Development Areas* 9(3): 335–338.

Collinson, M. P. 1972. *Farm Management in Peasant Agriculture: A Handbook for Rural Development Planning in Africa.* New York: Praeger.

Colombo, Umberto, D. Gale Johnson, and Toshio Shishido. 1977. "Expanding Food Production in Developing Countries: Rice Production in Southeast Asia." Trilateral Commission.

Colorado State University Field Party and Mona Reclamation Project Staff. 1977. *Watercourse Improvement in Pakistan: Pilot Study in Cooperation with Farmers at Tubewell 56L.* Water Management Technical Report No. 45. Fort Collins: Colorado State University.

Conlin, Sean. 1974. "Participation versus Expertise." *International Journal of Comparative Sociology,* 15:155–166.

Consultative Group on International Agricultural Research. 1976a. *Annual Report.* New York.

———. 1976b. *The Work of the Consultative Group on International Agricultural Research.* New York: CGIAR.

Contreras, Mario, et al. 1977. "An Interdisciplinary Approach to International Agricultural Training: The Cornell-CIMMYT Graduate Student Team Report." Ithaca: International Agriculture Program, Cornell University.

DeBoer, A. J., and A. Weisblat. 1980. *Livestock Component of Small Farm Systems in South and Southeast Asia.* Bellagio Conference on Integrated Crop and Animal Production, 1978. New York: Rockefeller Foundation.

de los Reyes, Romana P. 1980a. "Farmers' Approaches to Irrigation Management: Findings of a Study on the Management of 50 Philippine Communal Gravity Systems." Paper presented at International Rice Research Institute.

———. 1980b. *47 Communal Gravity Systems: Organization Profiles.* Quezon City: Institute of Philippine Culture.

Dennis, John V., Jr. 1978. "Buffalos vs. Tractors: A Comparison of Their Use in Village Agriculture of Northern Thailand." Paper presented at Cropping System Seminar, Khon Kaen University, Khon Kaen, Thailand.

Desowitz, Robert S., 1977. "The Fly That Would Be King." *Natural History* 76(2): 76–83.

Dillon, John L. 1971. "Interpreting Systems Simulation Output for Managerial Decision-Making." In J. B. Dent and J. R. Anderson, eds., *Systems Analysis in Agricultural Management,* pp. 85–120. Sydney: Wiley.

———. 1976. "The Economics of Systems Research." Agricultural Systems 1: 5–22.

———. 1979. "Broad Structural Review of the Small-Farmer Technology Problems." In Alberto Valdez, Grant M. Scobie, and John L. Dillon, eds., *Economics and the Design of Small-Farmer Technology.* Ames: Iowa State University Press.

Downs, Anthony. 1967. *Inside Bureaucracy.* Boston: Little, Brown.

Drosdoff, Matthew, ed. 1979. *World Food Issues.* Ithaca: Program in International Agriculture, Cornell University.

Dudal, R. 1978. "Land Resources for Agricultural Development." *Transactions of the 18th International Congress of Soil Science,* Plenary Papers, Volume 2, pp. 341–359. Edmondton, Canada: University of Alberta.

Duff, J. Bart. 1980. *The Potential for Mechanization in Small Farm Production Systems*. Bellagio Conference on Integrated Crop and Animal Production, 1978. New York: Rockefeller Foundation.

Erasmus, C. J. 1961. *Man Takes Control*. New York: Bobbs-Merrill.

Esman, Milton. 1978. *Landlessness and Near-Landlessness in Developing Countries*. Ithaca: Rural Development Committee, Cornell University.

Esman, Milton, Royal Colle, Norman Uphoff, and Ellen Taylor. 1980. *Paraprofessionals in Rural Development*. Ithaca: Rural Development Committee, Cornell University.

Esman, Milton, and Norman Uphoff. 1982. *Local Organization and Rural Development, The State of the Art*. Ithaca: Rural Development Committee, Cornell University.

FAO. 1969–1970. "Provisional Indicative World Plan for Agricultural Development. Prepared for 15th Session of the FAO Conference, C69M, Rome.

_____. 1978. *Report on the Agro-ecological Zones Project*. Volume 1: *Methodology and Results for Africa*. World Resources Report No. 48/1. Rome: FAO.

_____. 1979. *Production Yearbook*. Volume 31. Rome: FAO.

FAO/UNESCO. 1974. *Soil Maps of the World*. Volume 1 Legend. Paris: UNESCO.

Farmer, B. H., ed. 1977. *Green Revolution?: Technology and Change in Rice-Growing Areas of Tamil Nadu and Sri Lanka*. Boulder, Colo.: Westview.

Fernandez, Fernando. 1977. "Objectives and Content of Training at the International Centers of Agricultural Research." New York: Consultative Group on International Agricultural Research. Mimeo.

Fliegel, F. C., P. Roy, L. K. Sen, and J. K. Kivlin. 1968. *Agricultural Innovations in Indian Villages*. Hyderabad: National Institute of Community Development (India).

Fortmann, Louise. 1982. "Taking the Data Back to the Village." *Rural Development Participation Review* 3(2): 13–16.

Francis, C. A. 1978. "Multiple Cropping Potentials of Beans and Maize." *Horticulture Science* 13(1): 12–17.

Giles, G. W. 1975. "The Reorientation of Agricultural Mechanization for the Developing Countries: Policies and Attitudes for Action Programs." *Agricultural Mechanization in Asia* 6(2).

_____. 1967. "World Food Problems—Basic Needs." Annual Meeting, American Society of Agricultural Engineering, Paper no. 67–501.

Gladwin, Christina H. 1979. "Cognitive Strategies and Adoption Decisions: A Case Study of Nonadoption of an Agronomic Recommendation." *Economic Development and Cultural Change* 28(1): 155–173.

Goe, M. R., and R. E. McDowell. 1981. "Animal Traction: Guidelines for Utilization." Ithaca: International Agriculture, Cornell University. Mimeo.

Gostyla, Lynn, and William F. Whyte. 1980. *ICTA in Guatemala: The Evolution of a New Model for Agricultural Research and Development*. Ithaca: Rural Development Committee, Cornell University. Also published in Spanish as *El ICTA en Guatemala: La evolución de un modelo de investigación y desarrollo agrícolas*.

Green, James. 1961. "Success and Failure in Technical Assistance: A Case Study." *Human Organization* 20(1): 2–10.

Greenwood, Davydd J. 1973. *The Political Economy of Peasant Family Farming: Some Anthropological Perspectives on Rationality and Adaptation.* Ithaca: Rural Development Committee, Cornell University.

———. 1978. *Community-level Research, Local-Regional-Governmental Interactions and Development Planning: A Strategy for Baseline Studies.* Ithaca: Rural Development Committee, Cornell University.

Grigg, D. G. 1974. *The Agricultural Systems of the World: An Evolutionary Approach.* London: Cambridge University Press.

Grove, A. T., and F.M.C. Klein. 1979. *Rural Africa.* Cambridge: Cambridge University Press.

Gunkel, W. W. 1968. "Implementation and Over-Mechanization in Developing Countries." *Transactions of the American Society of Agricultural Engineering,* Paper no. 65–508.

Hansen, Gary. 1972. "Regional Administration for Rural Development in Indonesia: The Bimas Case." Working Paper Series 26. Honolulu: East-West Center.

Hargreaves, George H. 1972. *Deficiencias de aguas en Centro América y Panamá: Proyecto hidro-meteorológico Centro Americano.* Managua: UNDP World Meteorological Organization.

Harwood, Richard R. 1979. *Small Farm Development: Understanding and Improving Farming Systems in the Humid Tropics.* Boulder, Colo.: Westview.

Hayami, Yujiro. 1982. "Growth and Equity: Is There a Trade-Off?" Paper presented at International Conference on Agricultural Economics, Jakarta, Indonesia, August 24–September 2.

Hayami, Yujiro, and Vernon W. Ruttan. 1971. *Agricultural Development: An International Perspective.* Baltimore: John Hopkins Press.

Heginbotham, Stanley J. 1975. *Cultures in Conflict: The Four Faces of Indian Bureaucracy.* New York: Columbia University Press.

Hildebrand, Peter E. 1977. "Generating Small Farm Technology: An Integrated Multidisciplinary System." Paper presented at the 12th West Indian Agricultural Economics Conference. Antigua: Caribbean Agro-Economics Society.

———. 1978. Report to the Rockefeller Foundation on ICTA.

Hoffnar, Barnard R., and Glenn C. Johnson. 1966. *Summary and Evaluation of the Cooperative Agronomic-Economic Experimentation at Michigan State University, 1955–1963.* Michigan State University Research Bulletin #11. East Lansing.

Holdridge, L. R. 1967. *Life Zone Ecology.* With photographic supplements prepared by Joseph A. Tosi. Rev. ed. San José, Costa Rica: Tropical Science Center.

Howes, Michael, and Robert Chambers. 1979. "Indigenous Technical Knowledge: Analysis, Implications and Issues." In *Rural Development: Whose Knowledge Counts?* IDS Bulletin 10(2).

Inkeles, Alex. 1975. "Becoming Modern: Individual Change in Six Developing Countries." *Ethos* 3(2), 323–342.

Innes, Donald Q. 1980. "The Future of Traditional Agriculture." *Focus* 30(3): 1–3.

International Agricultural Development Service. 1979. "Preparing Professional Staff for National Agricultural Research Programs." *Bellagio Conference Workshop Report.*

IRRI. 1977. *Proceedings of Symposium on Cropping Systems Research and Development for the Asian Rice Farmer.* Los Baños, Philippines.

Isles, Carlos P., and Manuel L. Collado. 1979. "Farmer Participation in Communal Irrigation Development: Lessons from Laur." *Philippine Agricultural Engineering Journal* 10(2): 3–4 and 9.

Javier, E. Q. 1980. *Integration of Forages into Small Farming Systems.* Bellagio Conference on Integrated Crops and Animal Production, 1978. New York: Rockefeller Foundation.

Jensen, Einar, John W. Klein, Emil Rauchenstein, T. E. Woodward, and Ray H. Smith. 1942. *Input-Output Relationships in Milk Production.* Washington: U.S. Department of Agriculture Technical Bulletin 815.

Johnson, Ana Gutierrez. 1976. "Cooperativism and Justice: A Study and Cross-Cultural Comparison of Preferences for Forms of Equity among Basque Students of a Cooperative School-Factory." Masters Thesis, Cornell University.

Joosten, J. H. K. 1962. "Wirtschaftliche und agrarpolitische Aspekte Landbau Systema." Göttingen, Germany: Institut fuer Landwirtschaftliche Betriebslehre, 1962 (cited by Ruthenburg, 1971).

Kahl, Joseph. 1976. *Modernization, Exploitation and Dependency in Latin America: Germani, Gonzalez Casanova and Cardoso.* New Brunswick, N.J.: Transaction Books.

Kass, Donald C. L. 1978. *Polyculture Cropping Systems: Review and Analysis.* Ithaca: International Agriculture Bulletin 32, Cornell University.

Kikuchi, Masao, and Yujiro Hayami. 1980. "Inducements to Institutional Innovations in an Agrarian Community." *Economic Development and Cultural Change* 23: 21–36.

King, Franklin Hiram. 1911. *Farmers of Forty Centuries.* Madison, Wis.: Mrs. F. H. King. Reprinted 1973 by Rodale Press, Emmaus, Pa.

Kolars, J. F., and D. Bell. 1975. *Physical Geography—Environment and Man.* New York: McGraw-Hill.

Korten, David C. 1980. "Community Organization and Rural Development: A Learning Process Approach." *Public Administration Review* 40(5): 480–511.

——. 1981. "The Management of National and Social Transformation." *Public Administration Review* 4(6): 609–618.

Korten, David C., and Filipe B. Alfonso. 1981. *Bureaucracy and the Poor: Closing the Gap.* Singapore: McGraw-Hill.

Korten, Frances. 1981. "Community Participation: A Management Perspective on Obstacles and Options." In David C. Korten and Filipe B. Alfonso, *Bureaucracy and the Poor: Closing the Gap,* pp. 181–200. Singapore: McGraw-Hill.

——. 1982. "Building National Capacity to Develop Water Users' Associations: Experience from the Philippines." Staff Working Paper 528. Washington, D.C.: World Bank.

Koval, Andrew J., and Ahmed A. Behgat. 1980. "Ten Horsepower Agriculture." Catholic Relief Service Program, Arab Republic of Egypt. Paper presented at symposium on Appropriate Mechanization of Small Farms in Africa by Kenya National Academy for Advancement of Arts and Sciences and Association for Advancement of Agricultural Sciences in Africa, Nairobi.

Landsberg, H. E., et al. 1965. *World Maps of Climatology.* Edited by E. Rodenwaldt and H. J. Jusatz. Berlin: Springer.

———. 1969. *World Survey of Climatoloty.* Amsterdam: Elsevier.

Lassen, Cheryl. 1980a. *Landlessness and Rural Poverty in Latin America: Conditions, Trends and Policies Affecting Income and Employment.* Ithaca: Rural Development Committee, Cornell University.

———. 1980b. *Reaching the Assetless Poor: Projects and Strategies for Their Self Reliant Development.* Ithaca: Rural Development Committee, Cornell University.

Leagans, J. P. 1971. "Extension Education and Modernization." In J. P. Leagans and C. P. Loomis, eds., *Behavioral Change in Agriculture.* Ithaca: Cornell University Press.

Lele, Uma. 1975. *The Design of Rural Development: Lessons from East Africa.* Baltimore: Johns Hopkins University Press.

Leonard, David K. 1977. *Reaching the Peasant Farmer: Organization Theory and Practice in Kenya.* Chicago: University of Chicago Press.

Levine, Gilbert, and E. W. Coward. 1979. "Managing Water Resources for Food Production." Paper no. 4 in Matthew Drosoff, ed., *World Food Issues.* Ithaca: Program in International Agriculture, Cornell University.

Maner, J. H. 1980. *Non-Ruminants for Small Farm Systems.* Bellagio on Integrated Crop and Animal Production, 1978. New York: Rockefeller Foundation.

Martin, Lee, ed. 1977. *A Survey of Agricultural Economics Literature.* Volume 1: *Traditional Fields of Agricultural Economics, 1940's to 1970's.* Minneapolis: University of Minnesota Press.

McConnell, Grant. 1969. *The Decline of Agrarian Democracy.* New York: Atheneum.

McDowell, R. E. 1977. "Ruminant Products: More than Meat and Milk." Winrock Report No. 2. Morilton, Ark. Winrock International Livestock Research Training Center.

———. 1978. *Factors Limiting Animal Production in Small Farm Systems.* Bellagio Conference on Integrated Crop and Animal Production, 1978. New York: Rockefeller Foundation.

———. 1979. "Role of Animals in Support of Man." Paper no. 7 in Matthew Drosdoff, ed., *World Food Issues.* Ithaca: Program in International Agriculture, Cornell University.

———. 1980. "The Role of Animals in Developing Countries." In R. L. Baldwin, ed., *Animals, Feed, Food and People.* AAAS Selected Symposium No. 42. Boulder, Colo.: Westview.

———. 1981. "Relevant Technology Development and Dissemination from the International Agricultural Research Centers." Paper presented at the symposium:

Small Farms in a Changing World, Prospects for the 80's, Kansas State University, Manhattan.

McDowell, R. E., and P. E. Hildebrand. 1980. *Integrated Crop and Animal Production: Making the Most of Resources Available to Small Farms in Developing Countries.* Bellagio Conference on Integrated Crop and Animal Production, 1978. New York: Rockefeller Foundation.

Millikan, Max, and David Hapgood. 1966. *No Easy Harvest: The Dilemma of Agriculture in Developing Countries.* Boston: Little, Brown.

Montgomery, John D. 1957. *Forced to Be Free.* Chicago: University of Chicago Press, 1957.

Mosher, A. T. 1957. *Technical Cooperation in Latin American Agriculture.* Chicago: University of Chicago Press.

———. 1966. *Getting Agriculture Moving.* New York: Praeger.

———. 1969. *Creating a Progressive Rural Structure.* New York: Agricultural Development Council.

Norman, David W. 1973. *Methodology and Problems of Farm Management Investigations: Experiences from Northern Nigeria.* African Rural Employment Paper no. 8. East Lansing: Department of Agricultural Economics, Michigan State University.

———. 1980. "The Farming Systems Approach: Relevancy for Small Farmers." MSU Rural Development Paper no. 5. East Lansing: Department of Agricultural Economics, Michigan State University.

Núñez del Prado, Oscar. 1975. *Kuyo Chico: Experiment in Applied Anthropology.* Chicago: University of Chicago Press.

Okigbo, Bede N. 1978. "Cropping Systems and Related Research in Africa." Association for the Advancement of Agricultural Sciences in Africa, Occasional Publication Series OT.1.

Okigbo, Bede N., and D. J. Greenland. 1976. *Intercropping Systems in Tropical Africa: Multiple Cropping.* Madison, Wisc.: American Society for Agronomy.

Oram, Peter, et al. 1979. *Investment and Input Requirements for Accelerating Food Production in Low-Income Countries by 1990.* Research Report no. 10. Washington, D.C.: International Food Policy Research Institute.

Owen, Wyn R. 1971. *Two Rural Sectors: Their Characteristics and Roles in the Development Process.* Ottawa: International Development Research Centre.

Plucknett, D. L. 1979. *Managing Pastures and Cattle under Coconuts.* Westview Tropical Agriculture Series No. 2. Boulder, Colo.: Westview.

Porter, Keith S., 1975. *Nitrogen and Phosphorus Food Production: Waste and Environment.* Ann Arbor, Mich.: Ann Arbor Science Publishers.

President's Science Advisory Panel on the World Food Supply. 1967. *The World Food Problem.* Volume 2, Chapter 9.0, Tropical Soils and Climate. Washington, D.C.: U.S. Government Printing Office. 1967.

Propp, Kathleen M. 1968. *The Establishment of Agricultural Universities in India.* Special Publication 15. Urbana, Ill.: College of Agriculture, University of Illinois.

———. 1970. *A Method of Assessing Progress of Agricultural Universities in India.* New Delhi: Indian Council of Agricultural Research.

Recent State and Prospects of Agricultural Mechanization in Developing Countries. 1976. Proceedings of a Scientific Conference of the Institute of Tropical Agriculture and Veterinary Science, Karl-Marx-University of Leipzig, German Democratic Republic, Berlin: Akademie-Verlag.

Rice, E. B. 1971. *Extension in the Andes*. Washington, D.C.: Agency for International Development.

Rice, E. B., and E. Glaeser. 1972. *Agricultural Sector Studies, An Evaluation of AID's Recent Experience*. Washington: Agency for International Development, Evaluation Paper 5.

Riley, J. J. 1980. *Land, Water and Man as Determinants in Small Farm Production Systems*. Bellagio Conference on Integrated Crop and Animal Production, 1978. New York: Rockefeller Foundation.

Rockefeller Foundation. 1974, 1975. *Strategies for Agricultural Education in Developing Countries. Reports on First and Second Bellagio Conferences*. New York: Rockefeller Foundation.

Rogers, E. M. 1962. *Diffusion of Innovations*. New York: Free Press.

———. 1969. *Modernization among Peasants: The Impact of Communication*. New York: Holt, Rinehart, and Winston.

Rosenberg, David, and Jean Rosenberg. 1979a. *Landless Peasants and Rural Poverty in Selected Asian Countries*. Ithaca: Rural Development Committee, Cornell University.

———. 1979b. *Landless Peasants and Rural Poverty in Indonesia and the Philippines*. Ithaca: Rural Development Committee, Cornell University.

Ruano, Sergio. 1980. "Farming Systems Research in Guatemala: The ICTA Experience." Masters thesis, Cornell University.

Ruthenberg, Hans. 1971. *Farming Systems in the Tropics*. Oxford: Clarendon Press. 2d ed. 1976.

Saint, William S., and E. Walter Coward, Jr. 1977. "Agriculture and Behavioral Science: Emerging Orientations." *Science* 197: 733–737.

Schultz, Theodore W. 1964. *Transforming Traditional Agriculture*. New Haven: Yale University Press.

———, ed. 1978. *Distortions of Agricultural Incentives*. Bloomington: Indiana University Press.

Shrestha, Bihari K. 1980. "Nuwakot District (Nepal)." In *The Practice of Local-level Planning: Case Studies in Selected Rural Areas of India, Nepal, and Malaysia*. Bangkok: UN Economic and Social Commission for Asia and Pacific.

Sisler, Daniel G., and David R. Colman. 1979. *Poor Rural Households, Technical Change, and Income Distribution in Developing Countries: Insights from Asia*. Ithaca: Department of Agricultural Economics, Cornell University.

Soil Survey Staff, Soil Conservation Service. 1975. *Soil Taxonomy: A Basic System of Soil Classification for Making and Interpreting Soil Surveys*. U.S. Department of Agriculture, Handbook No. 436. Washington, D.C.: U.S. Government Printing Office.

Staudt, Kathleen. 1975. "Women Farmers and Inequities in Agricultural Service." *Rural Africana* 29: 81–94.

Stavis, Benedict. 1974. *Rural Local Governance and Agricultural Development in Taiwan*. Ithaca: China-Japan Program and Rural Development Committee, Cornell University.

Sussman, Gerald. 1980. "The Pilot Project and the Choice of an Implementation Strategy: Community Development in India." In Merilee S. Grindle, ed., *Politics and Policy Implementation in the Third World*. Princeton: Princeton University Press.

Swanson, Earl R. 1979. "Working with Other Disciplines." *American Journal of Agricultural Economics*, 61: 849.

Tabbal, D. F., and T. H. Wickham. 1978. "Effects of Location and Water Supply on Water Shortages in an Irrigated Area." In *Irrigation Policy and Management in Southeast Asia*, pp. 93–102. Los Baños, Philippines: International Rice Research Institute.

Turk, Kenneth L. 1980. *The Cornell-Los Baños Story*. Ithaca: New York State College of Agriculture and Life Sciences.

Turrent, Antonio. 1978. "El sistema agrícola: Un marco de referencia necessario para la investigación agrícola en Mexico." Unpublished.

Uphoff, Norman, and Milton Esman. 1974. *Local Organization for Rural Development: Analysis of Asian Experience*. Ithaca: Rural Development Committee, Cornell University.

Vander Velde, Edward J., et al. 1981. "Rehabilitating Gal Oya: Improving Water Management through Rural Participation in Sri Lanka." Paper presented at International Geographical Union, Fresno, California.

Wade, Robert. 1980. "India's Changing Strategy of Irrigation Development." In E. Walter Coward, Jr., ed., *Irrigation and Agricultural Development in Asia: Perspectives from the Social Sciences*. Ithaca: Cornell University Press.

Wade, Robert, and Robert Chambers. 1980. "Managing the Main System: Canal Irrigation's Blind Spot." *Economic and Political Weekly* 15(6): 39.

Weitz, Raanan. 1971. *From Peasant to Farmer: A Revolutionary Strategy for Development*. New York: Columbia University Press.

Whittlesey, D. 1976. "The Major Agricultural Regions of the World." *Annals of the Association of American Geographers* 26: 199–240.

Whyte, William F. 1964. "Culture, Industrial Relations, and Economic Development: The Case of Peru." *Industrial and Labor Relations Review* 16(4): 583–594.

———. 1975. *Organizing for Agricultural Development*. New Brunswick, N.J.: Transaction Books.

———. 1977. "Potatoes, Peasants and Professors: A Development Strategy for Peru. *Sociological Practice* 21(1): 7–23.

———. 1979. "On Making the Most of Participant Observation." *American Sociologist* 14: 55–66.

———. 1981. *Participatory Approaches to Agricultural Research and Development: A State of the Art Paper*. Ithaca: Rural Development Committee, Cornell University.

Whyte, William F., and Giorgio Alberti. 1976. *Power, Politics and Progress: Social Change in Rural Peru*. Amsterdam: Elsevier.

Whyte, William F., and L. K. Williams. 1968. *Toward an Integrated Theory of Development*. Ithaca: New York State School of Industrial and Labor Relations, Cornell University.

Williams, L. K., W. F. Whyte, and C. S. Green. 1966. "Do Cultural Differences Affect Workers' Attitudes?" *Industrial Relations* 5(3): 105–117.

Wolf, Eric. 1966. *Peasants*. Englewood Cliffs, N.J.: Prentice-Hall.

Wood, Alan. 1950. *The Groundnut Affair*. London: Bodley Head.

Wortman, Sterling, and Ralph W. Cummings, Jr. 1978. *To Feed This World: The Challenge and the Strategy*. Baltimore: Johns Hopkins University Press.

Zandstra, H. G., K. Swanberg, and C. A. Zulberti. 1976. *Removing Constraints to Small Farm Production: The Caqueza Project*. IDRC-058e. Ottawa: International Development Research Centre.

Zandstra, Hubert, Kenneth Swanberg, Carlos Zulberti, and Barry Nestel. 1979. *Caqueza: Living Rural Development*. Ottawa: Canada: International Development Research Centre.

Index

ANACH (peasant organization, Honduras), 180, 189, 202
Africa: animals in, 96–99; cropping systems in, 81–82, 96–99, 158; farming system in, 96–99; food staples of, 80; mechanization in, 114–115; research projects in, 34–36, 46–51, 110, 158, 258; U.S. models in, 18, 20. *See also* Ethiopia; Kenya; West Africa
Agencies, agricultural, redesign of: central working group in, 244–247; coordination, 235–238; decentralization, 234–235; to help farmers, 133–134; staffing, 242–244. *See also* Agricultural system, new; Bureaucracies, agricultural; Government programs
Agricultural extension, U.S.: agents, 35, 225; approach, 19; doctrine, 24; functions, 20; models, 21, 29
Agricultural system, new: approach, 41, 51–52; description, 11–15; integration of small farmers in, 39–40, 156–163, 233, 243, 312; learning process in, 244–248; new organizational models, 30, 164–193, 305–306; planning perspective, 231. *See also* Participatory system; Research and development strategy, new
Agronomists, in research projects, 171, 203
Ali, Hashim, 220
Animals, in farm systems, 90–110, 122; cultural importance, 107; importance of, 90–91, 124–125, 260;

research on, 161, 162; value, 105, 107, 125
Area coordinator, 237–238
Asia: farming systems, scheme, 94–95; humid-upland system, 103–105; irrigation development, 221; local organizations in, 207n; lowland rice system, 105; new crop research in, 157–158; shifting cultivation system, 101, 102, 103; U.S. models in, 18, 20; U.S. programs, 21. *See also* East Pakistan; India; Nepal; Pakistan, Southeast Asia; Taiwan
Asian Development Bank, 12, 297

BANDESA (Guatemala), 166
Bellagio Conferences: on educational programs for agricultural development (1974, 1975, 1979), 295–300; on farming systems including animals (1978), 90
Binswanger, Hans P., studies by, 113–114
Black, John D., 265
Borlaug, Norman, 27
Bradfield, Richard, 157; system, 157–158
Bradfield, Stillman, 160
Brazil: different soils of, 62, 64, 65
Bureaucracies, agricultural, 11, 133; conventional values in, 228–233; failures in, 225–226; patterns in, 227–228; political forces in, 222–224; problems for leaders, 224–225; reorientation in research and development, need for, 222,

Cultivation systems, major: fallow systems, 77–78, 79; field systems, 78, 79. *See also* Cropping systems; Farming systems

DIGESA (Dirección General de Servicios Agrícolas; Guatemala), 166; and ICTA, 165, 174, 176, 192; orientation, 174
Dennis, John V., studies by, 114
Developing countries. *See* LCDs
Development, agricultural: peasant participation in, 206–208; problems of, 133–134, 195. *See also* Research and development
Development programs, national: evaluation, 262–263; evolutionary approach, 261; involvement of farmers in, 252–253, 262–263; organization, 260–262
Dillon, John L., laws of simulation, 272
Dirección General de Servicios Agrícolas. *See* DIGESA

East Pakistan, research project in, 31–34, 46–51
Economic questions, for small farmers, 190; capital shortage, 198; and cooperatives, 203–204; credit cooperatives, 32–33; credit programs, 20, 28, 32–33, 35, 37, 42, 43, 184, 198; labor, 197; loans, 188; supplementary income, 138, 139, 197. *See also* Marketing
Economists, role in R&D, 198, 266, 277, 278
Educational programs, reorientation: conference on, 295, 298, 299; crucial need for, 309; formal study, 296–299; planning for, 302; problems, 296–298; regional centers, 299–300; training of professionals, 225–226, 295–300, 308–309; university's role, 306–308, 309
Ejido, 202, 203; definition, 202n
Environment, physical: factors in, 130–132; and farmer opportunities, 251; health problems, 70; impact of development on, 69; risks, 280. *See*

also Climate and ecology; Rainfall; Soils
Erosion, 81; problems of, 101, 103
Ethiopia: extension services in, 20; farmers in, 34–36, 49, 50; research project (CADU), 34–36, 46–51
Evapotranspiration: definition, 56n; effects, 56–58

FAO (Food and Agricultural Organization): assessment methods, 73; educational program, 298; studies by, 58–59, 73, 187, 258
FORUSA (Mexico) model: economics, 203–204; establishment, 202–203; limitations, 204–205, 206; program, 203–204; training programs, 301–302
Facilitator: assets, 290; definition, 286–287; knowledge needed by, 285; role, 288–289, 291; techniques, 288–289
Farm, small: description, 14, 25; efficiency and productivity of, 147, 195; experiments on, 160–163, 167–173, 193, 259, 294; stereotypes about, 310–311, 314
Farm family: importance of, 312; in LDCs, 14–15, 25; and labor, 132, 137, 138, 197–198; land as economic base, 139–140; role in decisions, 137, 279; and social system, 132–133; 148–149, 199; structure of, 138–139, 152
Farm laborers, landless, 13–14, 35
Farm Security Administration (FSA), 20–21
Farmer organization(s), 12–13; advantages, 317; building of, 196; cooperatives, 45, 151, 187, 203; economic significance, 316–317; and educational programs, 301–302; importance, 45, 178, 192, 201, 202, 208, 243–244; and irrigation, 216–217; need for, 195; political importance, 316; problems, 201
Farmer trials, description, 256–257
Farmers, larger: advantages for, 148, 152–153, 311–312, 317; involve-

United States (*cont.*)
culture, 18, 21–26, 155; organizational models, 18, 20, 28–30, 292–293; prestige and power of, 18; programs, 21. *See also* Agricultural extension, U.S.

Vellani, Rolando, 187–189
Villages, peasant: and market economy, 149–152; need for collective control, 283; and outsiders, 289–291; studies of, 139–145

Water resources: development, impact of, 69–71; management of, 67, 131; need for, 123. *See also* Irrigation
West Africa: cropping systems, 81, 85; research center in, 120
Wheat: development of, 157; as major crop, 99, 102, 312; research projects with, 34, 35, 48
Williams, Simon, experimental work by, 202
Women in farming, 136–138, 152
World Neighbors, project in Guatemala, 174, 175, 183

Zacapoaxtla (Mexico), project, 205

Library of Congress Cataloging in Publication Data

Main entry under title:

Higher-yielding human systems for agriculture.

  Includes bibliographical references and index.
   1. Agriculture—Research—Congresses. 2. Agriculture—
Social aspects—Research—Congresses. 3. Agricultural
systems—Research—Congresses. 4. Agriculture—Social
aspects—Congresses. 5. Agricultural systems—
Congresses. I. Whyte, William Foote, 1914–
II. Boynton, Damon.
S539.7.H54 1983   630   83-45151
ISBN 0-8014-1611-6 (alk. paper)